Statistics for Linguists:
An Introduction Using R

Statistics for Linguists: An Introduction Using R is the first statistics textbook on linear models for linguistics. The book covers simple uses of linear models through generalized models to more advanced approaches, maintaining its focus on conceptual issues and avoiding excessive mathematical details. It contains many applied examples using the R statistical programming environment. Written in an accessible tone and style, this text is the ideal main resource for graduate and advanced undergraduate students of Linguistics statistics courses as well as those in other fields, including Psychology, Cognitive Science, and Data Science.

Bodo Winter is Lecturer in Cognitive Linguistics in the Department of English Language and Applied Linguistics at the University of Birmingham, UK.

Statistics for Linguists: An Introduction Using R

Bodo Winter

Routledge
Taylor & Francis Group

NEW YORK AND LONDON

First published 2020
by Routledge
52 Vanderbilt Avenue, New York, NY 10017

and by Routledge
2 Park Square, Milton Park, Abingdon, Oxon, OX14 4RN

Routledge is an imprint of the Taylor & Francis Group, an informa business

Library of Congress Cataloging-in-Publication Data
Names: Winter, Bodo, author.
Title: Statistics for linguists : an introduction using R / by Bodo Winter.
Description: New York, NY : Routledge, 2019. | Includes bibliographical
    references and index.
Identifiers: LCCN 2019029350 (print) | LCCN 2019029351 (ebook) |
    ISBN 9781138056084 (hbk) | ISBN 9781138056091 (pbk) |
    ISBN 9781315165547 (ebk)
Subjects: LCSH: Linguistics—Statistical methods. | R (Computer program
    language) | Mathematical linguistics.
Classification: LCC P138.5 .W56 2019 (print) | LCC P138.5 (ebook) |
    DDC 410.1/5195—dc23
LC record available at https://lccn.loc.gov/2019029350
LC ebook record available at https://lccn.loc.gov/2019029351

ISBN: 978-1-138-05608-4 (hbk)
ISBN: 978-1-138-05609-1 (pbk)
ISBN: 978-1-315-16554-7 (ebk)

Typeset in Times New Roman
by Apex CoVantage, LLC

Contents

Acknowledgments

Let me take a few paragraphs to thank the people who have helped with this book. I've been lucky to have been exposed to some excellent statistics teaching. First, I want to thank Benjamin Bergen and Amy Schafer for getting me going with stats while I was a grad student at the University of Hawai'i at Mānoa. Second, I would like to thank Sarah Depaoli and Jack Vevea for teaching excellent graduate-level stats courses at the University of California, Merced. Finally, my biggest thanks go to Roger Mundry. I will never forget your workshops and our pizza nights.

Big thanks also go to Timo Roettger and Márton Sóskuthy for helping me develop the materials for the Birmingham Statistics for Linguists Summer School. I particularly want to thank Timo, a close friend and collaborator during all these years, for continuously challenging me. I hope that at some point I will become the scientist that lives up to his standards.

I want to thank Bruno Nicenboim for providing an excellent review of this book that led to many changes. I also want to thank Kamil Kaźmierski and Keith Wilson for additional suggestions. My student Greg Woodin has read all chapters and the book made a massive jump in quality thanks to his feedback. Another person who has been helping behind the scenes is my father Clive Winter, who has generously proofread early drafts of each chapter.

Special thanks go to the team at Routledge for their patience with me, as well as for their invaluable work on copy-editing and formatting this book (big shout-out for Nikky Twyman for her hard work on the manuscript). Sorry for being perennially late and not following the submission guidelines!

This book would not exist if not for Martine Grice, Anne Hermes, Doris Mücke, and Stefan Baumann. I taught my first course on R and mixed models for the phonetics group at the Institut für Linguistik.

I also want to thank all of the participants of the countless other workshops I have taught. If you were a member of one of my workshops, rest assured that it was your enthusiasm and your questions that allowed me to continuously refine the ways I explain certain concepts. I want to particularly thank the participants of the 2018 stats workshop at the Deafness Cognition and Language Research Centre at UCL, as well the participants of the first Birmingham Statistics for Linguists Summer School (2018).

I also want to thank those countless people who have sent me unsolicited thank-you emails in response to my freely available mixed model tutorial. Thanks for taking the time to reach out!

Finally, I want to thank the people who made life meaningful during the time I was writing this book, or who have supported me in other ways. These include my mum and dad, Mark, Louis, Vincenzo, Maciek, Marcus, Brittany, Jeannette, Matteo, Emily, Suganthi, Tommy, Dan, Jacob, Logan, Brendan, Jim, and Barry. I also want to include the Midlands Out badminton team, my yoga teachers Anna Robottom and Richard George, as well as the team of the Henrietta Street Gym for keeping me in high spirits.

0 Preface

Approach and How to Use This Book

The language sciences are undergoing a quantitative revolution. There is ever more data, an ever-growing toolkit of statistical methods, and a dedicated push towards incorporating more empirical research into linguistic theorizing. This book is designed to serve as a starting point for researchers from the language sciences and related disciplines to engage with emerging trends in data analysis. The plan is to take the reader from their first steps in R all the way to more advanced techniques such as linear mixed effects models. Along the way, the book aims to foster reproducible research practices.

Although this book is focused on the analysis of linguistic datasets, the examples chosen are accessible to researchers from other fields.

0.1. Strategy of the Book

This book is intended as a full course in basic linear modeling, from descriptive statistics, regression, and multiple regression, over to logistic regression and Poisson regression, all the way up to mixed models. Other books introduce statistics with more 'traditional' methods, such as t-tests, Chi-Square tests, ANOVAs, etc. I believe these significance tests are the wrong starting point for learners. When students are introduced to this 'testing framework', they spend most of their time worrying about what test to pick, rather than worrying about how to implement their theoretical understanding of a phenomenon in a statistical model. This book fosters model-based thinking rather than test-based thinking.

I have found out through experience that introducing learners to statistics via the linear model framework is much more engaging than teaching an array of significance tests. When starting one's statistical journey with these traditional methods, statistics seems like a long vocabulary list, and the student is left hanging with a highly compartmentalized view of the field. Moreover, learning statistics via significance tests gives all the wrong incentives. Significance tests such as t-tests, Chi-Square tests, and ANOVAs provide 'quick fixes' that encourage the researcher to treat the p-value as the ultimate arbiter of truth. Instead, students should be encouraged to think deeply about their theories, and they should be encouraged to spend a lot of time interpreting their models in substantive terms.

This book does not focus on the underlying mathematics, for which more advanced textbooks are available. The tone is deliberately casual, trying to make statistics approachable. Everything is geared towards practical relevance, aimed at researchers and students who want to model their data statistically to address their research questions. Some might argue that it is dangerous to teach linear models and their extensions without delving deeply into the underlying mathematics. Indeed, for every chapter in this book, there are book-length treatments that go into much more detail. So, naturally, I've had to cut corners somewhere in assembling this material. I do not intend to further statistical ignorance. Instead, I believe that there is value in a friendly, practical introduction geared towards usability, with more interested and mathematically capable readers being free to read more advanced texts further down the line.

There are many problems with how statistical methods are applied within the language sciences. For example, people use significance tests and *p*-values without knowing what they mean; people misinterpret main effects in the presence of interactions; and people fit models without being aware of convergence issues. This book is very much written with these problems in mind, based on my experience of teaching workshops to linguists at various institutions. Most of the errors that I see in the context of mixed models, for example, have to do with an insufficient understanding of the underlying regression framework. This is another reason for why this book focuses so much on linear models.

I think that part of the problem the field has with statistics stems from the fact that methods that are too advanced have been given to linguists too early. Baayen's landmark textbook *Analyzing Linguistic Data* (2008) ushered in a new quantitative era in linguistics. However, what many people end up using in practice is a pale reflection of what's discussed in Baayen's book or other texts. This state of affairs results from the fact that there is a scarcity of easy textbooks that provide stepping stones towards more advanced reads. In believe that, as a scientific community, we further bad statistics if we only write textbooks written for mathematically advanced readers.

This book is written in a friendly tone, directly addressing the reader ('you'), and explaining each formula and R function in quite a lot of detail. A lot of the datasets analyzed come from my own work because it allows me to guide the student through some of the reasoning processes that were going on behind the scenes. Altogether, I hope that this is a very relatable and non-threatening introduction to statistics with R.

Finally, this book tries to foster reproducible research practices. For example, the book gives advice about data sharing and reproducible code. Moreover, I will emphasize that a publication without concomitant release of the data and code has to be considered incomplete—unless there are very good reasons for not sharing these materials. Reproducible research takes effort and needs to be trained. This book tries to give the right incentives in these matters.

0.2. Why R?

This book is entirely based on the R programming language. These days, it's safe to say that R is the de facto standard in the language sciences. If you are reading this, chances are that the following paragraphs involve preaching to the choir, since picking up this book likely means that you are convinced of the necessity of using R.

However, since there are still many linguistics and psychology departments that teach their students proprietary software, in particular SPSS, it is worth highlighting why there is absolutely no way around R these days, and why this book *had* to be structured around R.

First, let me make a strong claim to begin with: if you are teaching your students SPSS rather than a programming language, such as R, you are actively disadvantaging their careers, as well as their success as scientists. Not only are there more job posts that ask for R skills than SPSS skills, but R is also much more conducive to open and reproducible research practices, which are required by an increasing number of academic journals and funding bodies. At some point in the near future, it will be difficult for your students to publish if they don't offer their data and code, a practice that the point-and-click structure of SPSS does not actively incentivize. I'd go as far as saying that, at this stage, teaching SPSS to students is unethical, because doing so is inherently directed against the open and community-driven nature of science.

Sometimes I hear the argument that R may be too difficult for students, in particular for undergraduates. In stark contrast to this, I've found that students from all sorts of backgrounds (even without any programming knowledge) can quickly pick up R if it is taught in a friendly manner. Moreover, it helps students that R can be used on their own machines without licensing hassle, and it further helps students that there's by now much more online help for R than for SPSS. Also, the interactive nature of R, as well as the ease with which plots can be created, can be highly engaging to students.

A final point about R is that it allows making 'data wrangling' an integral part of a statistical analysis. Because preprocessing the data and statistical modeling are two sides of the same coin, they should happen within the same software environment. R makes this easier than other software.

0.3. Why the Tidyverse?

These days, there are two 'dialects' or 'styles' of programming in R. One uses mostly 'base R' functions (those that come with the original R distribution). The other dialect uses 'tidyverse' packages. The 'tidyverse' is a label used for the network of packages developed by Hadley Wickham and colleagues, including such widely known packages as `dplyr` and `ggplot2`. Which style should you learn?

Essentially, there's no way around knowing *both* of these styles. A solid foundation in base R is still crucial, even if many tidyverse functions provide easier alternatives. Many web tutorials or discussions in online help forums such as StackOverflow include extensive base R code, but the student will invariably also encounter tidyverse-style code. Given this state of affairs, I think that it is necessary to teach both styles.

That said, the 'tidy' style is much easier to read and I've found that students grasp it more quickly. So I decided that there should be one introductory chapter on base R (Chapter 1), as well as one on the tidyverse (Chapter 2). However, after Chapter 2, the book almost exclusively uses tidyverse-style code from Chapter 2 onwards. Only when base R offers the easier alternative is base R code used.

0.4. R Packages Required for This Book

You need to download and install R and RStudio, which can be downloaded online. The following R packages need to be installed to be able to execute all code in all chapters. The tidyverse and broom packages should be loaded for every chapter, as they are used throughout the entire book.

```
install.packages('tidyverse')
install.packages('broom')
install.packages('gridExtra')
install.packages('car')
install.packages('MASS')
install.packages('pscl')
install.packages('effsize')
install.packages('lme4')
install.packages('afex')
install.packages('brms')
install.packages('MuMIn')
install.packages('swirl')
install.packages('languageR')
install.packages('emmeans')
```

0.5. What This Book Is Not

To get any false expectations out of the way, let me tell you a few things that this book is *not*.

*   This book is not an introduction to the underlying theory and mathematics of regression, or mixed models. For this, there are more advanced materials available. Beware that any introductory text will have to cut corners on some topics, and this one is no exception.
*   This book is not an introduction to exploratory techniques, such as exploratory factor analysis, cluster analysis, or classification and regression trees.
*   This book is not a 'cookbook' that teaches you a whole range of different techniques. The focus is on regression modeling. Appendix A shows how some basic significance tests (such as *t*-tests) map onto the techniques discussed throughout the book. The concluding chapter of this book, Chapter 16, provides additional arguments why the cookbook approach is limiting students, and why it should be avoided whenever possible.

0.6. How to Use This Book

The materials presented here are intended as a full course. Each chapter combines conceptual introductions to statistical topics with hands-on applications. To maximize learning benefits, it is of utmost importance that you actually execute the R code presented in each chapter. Only by typing in each and every command can you develop the relevant muscle memory to learn the programming language.

All the data that is needed for this book can be accessed via the following Open Science Framework (OSF) repository:

> https://osf.io/34mq9/

Some further recommendations:

- Although I do provide script files for all chapters on this webpage, I don't recommend looking at these while reading the book. Only consult these when you get stuck.
- The data can be downloaded from the website. It's possible to work through the entire book in one continued R session. Alternatively, you can also quit R after each chapter and come back to where you finished.
- I highly recommend setting up a folder on your computer where all the materials are saved, and where you create scripts that follow the code presented in each chapter. Annotate the code with your own comments to make it 'your own'.
- There are exercises at the end of each chapter. The solutions to the exercises can also be found on the above-mentioned repository.

0.7. Information for Teachers

This book is intended to be read from front to back. However, Appendix A (on significance tests) and Chapters 9 and 10 can be moved to different points, depending on the needs of a particular course. Likewise, Chapter 16 can be read independently of the other chapters, as well. This book can be used for both undergraduate and postgraduate courses. Chapter 8 on interactions is hard and, if teaching an undergraduate class, I may forestall this chapter at the expense of having more time to discuss inferential statistics (Chapters 9, 10, and 11).

If you've already taught statistics classes, you may be used to teaching a class that is focused on significance tests. In this case, I welcome you to consider the approach adopted in this book. Trust me, it works.

That said, you may want to continue teaching significance tests. In this case, this book could still be a useful textbook for your class, as the issues discussed here also apply to significance tests. Moreover, the methods discussed throughout the chapters have direct correspondences to significance tests, and these correspondences are explained in Appendix A.

All in all, I hope that this book strikes a nice balance between the easy and the advanced, so that readers from all levels will find something useful in it.

1 Introduction to R

1.1. Introduction

Statistics, conceived broadly, is the process of "getting meaning from data".[1] We perform statistical analyses on datasets to further our understanding. As such, statistics is a fundamentally human process that makes large sets of numbers embedded in complex datasets amenable to human cognition.

Some people think of statistics as being only the very last step of the empirical process. You design your study, you collect your data, *then* you perform a statistical analysis of the data. This is a narrow view of statistics.

This book assumes a broad view. In particular, I view the process of getting the data in shape for an analysis as part of your actual analysis. Thus, what people talk of as 'preprocessing' or 'data wrangling' is an integral part of statistics. In fact, most of your time during an actual analysis will be spent on wrangling with the data. The first two chapters focus on teaching you the skills for doing this. I will also teach you the first steps towards an efficient workflow, as well as how to do data processing in a reproducible fashion.

R makes all of this easy—once you get used to it. There's absolutely no way around a command-line-based tool if you want to be efficient with data. You need the ability to type in programming commands, rather than dealing with the data exclusively via some graphical user interface. Using a point-and-click-based software tool such as Excel slows you down and is prone to error. More importantly, it makes it more difficult for others to reproduce your analysis. Pointing and clicking yourself through some interface means that another researcher will have a difficult time tracing your steps. You need a record of your analysis in the form of programming code.

As a telling example of what can go wrong with processing your data, consider the case of 'austerity's spreadsheet error', which has been widely covered in the news:[2] Reinhart and Rogoff (2010) published an influential paper which showed that, on average, economic growth was diminished when a country's debt exceeds a certain limit. Many policy makers used this finding as an argument for austerity politics. However, the MIT graduate student Thomas Herndon discovered that the results were

---

1 This phrase is used by Michael Starbird in his introduction to statistics for *The Great Courses*.
2 For example [accessed, October 12, 2018]:
    www.theguardian.com/politics/2013/apr/18/uncovered-error-george-osborne-austerity
    www.bbc.co.uk/news/magazine-22223190
    www.aeaweb.org/articles?id=10.1257/aer.100.2.573

based on a spreadsheet error: certain rows were accidentally omitted in their analysis. Including these rows led to drastically different results, with different implications for policy makers. The European Spreadsheet Risks Interest Group curates a long list of spreadsheet "horror stories".[3] The length of this list is testament to the fact that it is difficult *not* to make errors when using software such as Excel.

The upshot of this discussion is that there's no way around learning a bit of R. The fact that this involves typing in commands rather than clicking yourself through some graphical interface may at first sight seem daunting. But don't panic—this book will be your guide. Those readers who are already experienced with R may skim through the next two chapters or skip them altogether.

1.2. Baby Steps: Simple Math with R

You should have installed R and RStudio by now. R is the actual programming language that you will use throughout this book. RStudio makes managing projects easier, thus facilitating your workflow. However, it is R embedded within RStudio that is the actual workhorse of your analysis.

When you open up RStudio, the first thing you see is the console, which is your window into the world of R. The console is where you type in commands, which R will then execute. Inside the console, the command line starts with the symbol '>'. Next to it, you will see a blinking cursor ' | '. The blinking is R's way of telling you that it's ready for you to enter some commands.

One way to think about R is that it's just an overblown calculator. Type in '2 + 2' and press ENTER:

```
2 + 2
```

```
[1] 4
```

This is addition. What about subtraction?

```
3 - 2
```

```
[1] 1
```

What happens if you supply an incomplete command, such as '3 −', and then hit ENTER? The console displays a plus sign. Hitting ENTER multiple times yields even more plus signs.

```
3 -
```

```
+
+
+
+
```

You are stuck. In this context, the plus sign has nothing to do with addition. It's R's way of showing you that the last command is incomplete. There are two ways out of this: either supplying the second number, or aborting by pressing ESC. Remember this for whenever you see a plus sign instead of a '>' in the console.

When you are in the console and the cursor is blinking, you can press the up and down arrows to toggle through the history of executed commands. This may save you some typing in case you want to re-execute a command.

Let's do some division, some multiplication and taking a number to the power of another number:

```
3 / 2 # division
```

```
[1] 1.5
```

```
3 * 2 # multiplication
```

```
[1] 6
```

```
2 ^ 2 # two squared
```

```
[1] 4
```

```
2 ^ 3 # two to the power of three
```

```
[1] 8
```

You can stack mathematical operations and use brackets to overcome the default order of operations. Let's compare the output of the following two commands:

```
(2 + 3) * 3
```

```
[1] 15
```

```
2 + (3 * 3)
```

```
[1] 11
```

The first is $2 + 3 = 5$, multiplied by 3, which yields 15. The second is $3 * 3 = 9$ plus 2, which yields 11. Simple mathematical operations have the structure 'A operation B', just as in mathematics. However, most 'functions' in R look different from that. The general structure of an R function is as follows:

    function(argument1, argument2, ...)

A function can be thought of as a verb, or an action. Arguments are the inputs to functions—they are what functions act on. Most functions have at least one argument.

If a function has multiple arguments, they are separated by commas. Some arguments are obligatory (the function won't run without being supplied a specific argument). Other arguments are optional.

This is all quite abstract, so let's demonstrate this with the square root function `sqrt()`:

```
sqrt(4)
```

```
[1] 2
```

This function only has one obligatory argument. It needs a number to take the square root of. If you fail to supply the corresponding argument, you will get an error message.

```
sqrt()
```

```
Error in sqrt() : 0 arguments passed to 'sqrt' which
requires 1
```

Another example of a simple function is the absolute value function `abs()`. This function makes negative numbers positive and leaves positive numbers unchanged, as demonstrated by the following two examples.

```
abs(-2)
```

```
[1] 2
```

```
abs(2)
```

```
[1] 2
```

1.3. Your First R Script

So far, you have typed things straight into the console. However, this is exactly what you *don't* want to do in an actual analysis. Instead, you prepare an R script, which contains everything needed to reproduce your analysis. The file extension .R is used for script files. Go to RStudio and click on 'File' in the menu tab, then click on 'New File' in the pop-down menu, then 'R Script'.

Once you have opened up a new script file, your RStudio screen is split into two halves. The top half is your R script; the bottom half is the console. Think of your R script as the recipe, and the R console as the kitchen that cooks according to your recipe. An alternative metaphor is that your R script is the steering wheel, and the console is the engine.

Type in the following command into your R script (*not* into the console) and press ENTER.

```
2 * 3
```

Nothing happens. The above command is only in your script—it hasn't been executed yet. To make something happen, you need to position your cursor in the line of the command and press the green arrow in the top right of the R script window. This will 'send' the instructions from the script down to the console, where R will execute the command. However, rather than using your mouse to click the green button, I strongly encourage you to learn the keyboard shortcut for running a command, which is COMMAND + ENTER on a Mac and ALT + ENTER on a PC.

When working with R, try to work as much as possible in the script, which should contain everything that is needed to reproduce your analysis. Scripts also allow you to comment your code, which you can do with the hashtag symbol '#'. Everything to the right of the hashtag will be ignored by R. Here's how you could comment the above command:

```
# Let's multiply two times three:
2 * 3
```

Alternatively, you can have the comment in the same line. Everything up to the comment is executed; everything after the comment is ignored.

```
2 * 3  # Multiply two times three
```

Commenting is crucial. Imagine getting back to your analysis after a two-year break, which happens surprisingly often. For a reasonably complex data analysis, it may take you hours to figure out what's going on. Your future self will thank you for leaving helpful comments. Perhaps even more importantly, comprehensive commenting facilitates reproducibility since other researchers will have an easier time reading your code.

Different coders have different approaches, but I developed the personal habit of 'commenting before coding'. That is, I write what I am going to do next in plain language such as '# Load in data:'. I supply the corresponding R code only after having written the comment. As a result of this practice, my scripts are annotated from the get-go. It also helps my thinking, because each comment states a clear goal before I start hacking away at the keyboard.

Maybe you don't want to adopt this habit, and that's OK. Programming styles are deeply personal and it will take time to figure out what works for you. However, regardless of how you write your scripts, write them with a future audience in mind.

1.4. Assigning Variables

Let's use the R script to assign variables. Write the following command into your R script (not into the console).

```
x <- 2 * 3
```

If you send this command in the console (remember: COMMAND + ENTER or CTRL + ENTER), nothing happens. The leftwards pointing arrow '<-' is the 'assign operator'. It assigns whatever is to the right of the operator to an 'object' that bears the name on the left. You decide the name yourself. In this case, the object is called x, and it stores the output of the operation '2 * 3'. I like to think of the arrow '<−' as metaphorically putting something into a container. Imagine a container with 'x' written on it that contains the number 6. By typing in the container's name, you retrieve its content.

```
x
```

```
[1]  6
```

You will also see code that uses a different assignment operator, namely '=' as opposed to '<−'.

```
x = 2 * 3
```

```
[1]  6
```

There is a subtle difference between the two assignment operators that I won't go into here. For now, it's best if you stick to '<−', which is also what most R style guides recommend. As you will be using the '<−' assign operator constantly, make sure to learn its shortcut: ALT + minus.

The x can be used in further mathematical operations as if it's a number.

```
x / 2
```

```
[1]  3
```

Crucially, R is case-sensitive. Typing in capital 'X' yields an error message because the object capital 'X' does not exist in your 'working environment'.

```
X
```

```
Error: object 'X' not found
```

To retrieve a list of all objects in your current working environment, type ls() (this function's name stands for 'list').

```
ls()
```

```
[1]  "x"
```

Since you just started a new session and only defined one object, your working environment only contains the object x. Notice one curiosity about the ls() function: it is one of the few functions in R that doesn't need any arguments, which is why running the function without an argument didn't produce an error message.

1.5. Numeric Vectors

Up to this point, the object x only contained one number. One handy function to create multi-number objects is the concatenate function c(). The following code uses this function to put the numbers 2.3, 1, and 5 into one object. As before, typing the object's name reveals its content. This command overrides the previous x.

```
x <- c(2.3, 1, 5)

x
```

```
[1] 2.3 1.0 5.0
```

The object x is what is called a vector. In R, a 'vector' simply is a list of numbers. You can check how long a vector is with the length() function.

```
length(x)
```

```
[1] 3
```

As you will see shortly, there are different types of vectors. The vector x contains numbers, so it is a vector of type 'numeric'. The mode() and class() function can be used to assess vector types.[4] People often either talk of a vector's 'atomic mode' or 'atomic class'.

```
mode(x)
```

```
[1] "numeric"
```

```
class(x)
```

```
[1] "numeric"
```

It's important to know what type of vector you are dealing with, as certain mathematical operations can only be applied to numeric vectors.

Let's create a sequence of integers from 10 to 1 using the colon function. In R, the colon is a sequence operator that creates an integer sequence from the first number to the last number.

```
mynums <- 10:1

mynums
```

```
[1] 10 9 8 7 6 5 4 3 2 1
```

---

4 These functions are equivalent for simple vectors, but exhibit different behavior for more complex objects (not covered here).

Given that mynums is a numeric vector, it is possible to perform all kinds of new mathematical operations on it. The following code showcases some useful summary functions.

```
sum(mynums)  # sum
```
```
[1] 55
```

```
min(mynums)  # smallest value (minimum)
```
```
[1] 1
```

```
max(mynums)  # largest value (maximum)
```
```
[1] 10
```

```
range(mynums)  # minimum and maximum together
```
```
[1] 1 10
```

```
diff(range(mynums))  # range: difference between min and max
```
```
[1] 9
```

```
mean(mynums) # arithmetic mean, see Ch. 3
```
```
[1] 5.5
```

```
sd(mynums)  # standard deviation, see Ch. 3
```
```
[1] 3.02765
```

```
median(mynums)     # median, see Ch. 3
```
```
[1] 5.5
```

If you use a function such as subtraction or division on a numeric vector, the function is repeated for all entries of the vector.

```
mynums - 5  # subtract 5 from every number
```
```
[1] 5 4 3 2 1 0 -1 -2 -3 -4
```

```
mynums / 2  # divide every number by two
```
```
[1] 5.0 4.5 4.0 3.5 3.0 2.5 2.0 1.5 1.0 0.5
```

1.6. Indexing

Often, you need to operate on specific subsets of data. Vectors can be indexed by position. Conceptually, it is important to separate a vector's position from the value that's stored at said position. Because each vector in R is ordered, it is possible to use indices to ask for the first data point, the second data point, and so on.

```
mynums[1]    # retrieve value at first position
```

```
[1] 10
```

```
mynums[2]    # retrieve value at second position
```

```
[1] 9
```

```
mynums[1:4] # retrieve first four values
```

```
[1] 10 9 8 7
```

Putting a minus in front of an index spits out everything inside a vector *except for that index.*

```
mynums[-2]   # retrieve everything except second position
```

```
[1] 10 8 7 6 5 4 3 2 1
```

Now that you know the basic of indexing, you can also understand why there's a '[1]' in front of each of line of R output you've seen so far. Creating a longer integer sequence will help you wrap your head around this.

```
1:100
```

```
 [1]   1   2   3   4   5   6   7   8   9  10  11  12  13  14
[15]  15  16  17  18  19  20  21  22  23  24  25  26  27  28
[29]  29  30  31  32  33  34  35  36  37  38  39  40  41  42
[43]  43  44  45  46  47  48  49  50  51  52  53  54  55  56
[57]  57  58  59  60  61  62  63  64  65  66  67  68  69  70
[71]  71  72  73  74  75  76  77  78  79  80  81  82  83  84
[85]  85  86  87  88  89  90  91  92  93  94  95  96  97  98
[99]  99 100
```

The '[1]' simply means 'first position'. Whenever there's a line break, R will show the position of the first value that starts a new row. The numbers in the square brackets to the left may be different on your screen: this will depend on your screen resolution and the resolution of the font printed in the R console.

1.7. Logical Vectors

Calling data by position is impractical for large datasets. If you had, for example, a dataset with 10,000 rows, you wouldn't necessarily know in advance that a particular data point of interest is at the 7,384th position. You need to be able to ask for specific values, rather than having to know the position in advance. For this, logical statements are useful.

```
mynums > 3   # Which values are larger than 3?
```

```
[1] TRUE TRUE TRUE TRUE TRUE TRUE
[7] TRUE FALSE FALSE FALSE
```

The statement mynums > 3 uses the 'greater than' sign '>'. This line of code is essentially the same as asking: 'Is it the case that mynums is larger than 3?' Because the vector mynums contains multiple entries, this question is repeated for each position, each time returning a TRUE value if the number is actually larger than 3, or a FALSE if the number is smaller than 3.

The logical operator '>=' translates to 'larger than or equal to'. The operators '<' and '<=' mean 'smaller than' and 'smaller than or equal to'. Have a look at what the following commands do, keeping in mind that the mynums vector contains the integer sequence 10:1.

```
mynums >= 3   # Larger than or equal to 3?
```

```
[1] TRUE TRUE TRUE TRUE TRUE TRUE
[7] TRUE TRUE FALSE FALSE
```

```
mynums < 4   # Smaller than 4?
```

```
[1] FALSE FALSE FALSE FALSE FALSE FALSE
[7] FALSE TRUE TRUE TRUE
```

```
mynums <= 4   # Smaller than or equal to 4?
```

```
[1] FALSE FALSE FALSE FALSE FALSE FALSE
[7] TRUE TRUE TRUE TRUE
```

```
mynums == 4   # Equal to 4?
```

```
[1] FALSE FALSE FALSE FALSE FALSE FALSE
[7] TRUE FALSE FALSE FALSE
```

```
mynums != 4   # Not equal to 4?
```

```
[1] TRUE TRUE TRUE TRUE TRUE TRUE
[7] FALSE TRUE TRUE TRUE
```

The result of performing a logical operation is actually a vector itself. To illustrate this, the following code stores the output of a logical operation in the object `mylog`. The `class()` function shows that `mylog` is 'logical'.

```
mylog <- mynums >= 3

class(mylog)
```

```
[1] "logical"
```

Logical vectors can be used for indexing. The following code only returns those values that are larger than or equal to 3.

```
mynums[mylog]
```

```
[1] 10 9 8 7 6 5 4 3
```

Perhaps it is more transparent to put everything into one line of code rather than defining separate vectors:

```
mynums[mynums >= 3]
```

```
[1] 10 9 8 7 6 5 4 3
```

It may help to paraphrase this command as if directly talking to R: 'Of the vector `mynums`, please retrieve those numbers for which the statement `mynums >= 3` is TRUE'.

1.8. Character Vectors

Almost all analysis projects involve some vectors that contain text, such information about a participant's age, gender, dialect, and so on. For this, 'character' vectors are used.

The code below uses quotation marks to tell R that the labels `'F'` and `'M'` are character strings rather than object names or numbers. You can use either single quotes or double quotes, but you should not mix the two.[5]

```
gender <- c('F', 'M', 'M', 'F', 'F')
```

The character-nature of the `gender` vector is revealed when printing the vector into the console, which shows quotation marks.

---

5 I use single quotes because it makes the code look 'lighter' than double quotes, and because it saves me one additional key stroke on my keyboard.

```
gender
```

```
[1] "F" "M" "M" "F" "F"
```

As before, the type of vector can be verified with `class()`.

```
class(gender)
```

```
[1] "character"
```

As before, you can index this vector by position or using logical statements.

```
gender[2]
```

```
[1] "M"
```

```
gender[gender == 'F']
```

```
[1] "F" "F" "F"
```

However, it is impossible to perform mathematical functions on this vector, and doing so will spit out a warning message.

```
mean(gender)
```

```
[1] NA
Warning message:
In mean.default(gender) : argument is not numeric or logi-
cal: returning NA
```

1.9. Factor Vectors

A fourth common type of vector is the 'factor' vector. The following code overrides the original `gender` vector with a new version that has been converted to a factor using `as.factor()`.[6]

```
gender <- as.factor(gender)
```

```
gender
```

```
[1] F M M F F
Levels: F M
```

The output shows text, but, unlike the character vector, there are no quotation marks. The 'levels' listed below the factor are the unique categories in the vector. In

---

6 There are also `as.numeric()`, `as.logical()`, and `as.character()`. Perhaps play around with these functions to see what happens (and what can go wrong) when you convert a vector of one type into another type. For example, what happens when you apply `as.numeric()` to a logical vector? (This may actually be useful in some circumstances.)

this case, the vector `gender` contains 5 data points, which are all tokens of the types "F" and "M". The levels can be accessed like this:

```
levels(gender)
```

```
[1] "F" "M"
```

The issue with factor vectors is that the levels are fixed. Let's see what happens when you attempt to insert a new value 'not_declared' into the third position of the `gender` vector.

```
gender[3] <- 'not_declared'
```

```
Warning message:
In '[<-.factor'('*tmp*', 3, value = "not_declared") :
 invalid factor level, NA generated
```

```
gender
```

```
[1] F   M   <NA> F   F
Levels: F M
```

The third position is now set to NA, a missing value. This happened because the only two levels allowed are 'F' and 'M'. To insert the new value 'not_declared', you first need to change the levels.

```
levels(gender) <- c('F', 'M', 'not_declared')
```

Let's re-execute the insertion statement.

```
gender[3] <- 'not_declared'
```

This time around, there's no error message, because 'not_declared' is now a valid level of the `gender` vector. Let's check whether the assignment operation achieved the expected outcome:

```
gender
```

```
[1] M            F           not_declared M            M
Levels: M F not_declared
```

1.10. Data Frames

Data frames are basically R's version of a spreadsheet. A data frame is a two-dimensional object, with rows and columns. Each column contains a vector.

Let's build a data frame. The following command concatenates three names into one vector.

```
participant <- c('louis', 'paula', 'vincenzo')
```

Next, the data.frame() function is used to create a data frame by hand. Each argument of this function becomes a column. Here, the participant vector will be the first column. The second column is named score, and a vector of three numbers is supplied.

```
mydf <- data.frame(participant, score = c(67, 85, 32))

mydf
```

```
  participant score
1    louis       67
2    paula       85
3 vincenzo       32
```

Because a data frame is two-dimensional, you can ask for the number of rows or columns.

```
nrow(mydf)
```

```
[1] 3
```

```
ncol(mydf)
```

```
[1] 2
```

The column names can be retrieved like this:

```
colnames(mydf)
```

```
[1] "participant" "score"
```

Data frames can be indexed via the name of the column by using the dollar sign operator '$'.

```
mydf$score
```

```
[1] 67 85 32
```

This results in a numeric vector. You can then apply summary functions to this vector, such as computing the mean:

```
mean(mydf$score)
```

```
[1] 61.33333
```

You can check the structure of the data frame with the str() function.

```
str(mydf)
```

```
'data.frame':  3 obs. of 2 variables:
$ participant: Factor w/ 3 levels "louis","paula",..: 1 2 3
$ resp       : num 67 85 32
```

This function lists all the columns and their vector types. Notice one curiosity: the `participant` column is indicated to be a factor vector, even though you only supplied a character vector! The `data.frame()` function secretly converted your character vector into factor vector.

The `summary()` function provides a useful summary, listing the number of data points for each participant, as well as what is called a 'five number summary' of the `score` column (see Chapter 3).

```
summary(mydf)
```

```
 participant    score
louis   :1   Min.   :32.00
paula   :1   1st Qu.:49.50
vincenzo:1   Median :67.00
             Mean   :61.33
             3rd Qu.:76.00
             Max.   :85.00
```

To index rows or columns by position, you can use square brackets. However, due to data frames being two-dimensional, this time around you need to supply identifiers for rows and columns, which are separated by comma, with rows listed first.

```
mydf[1,]  # first row
```

```
 participant score
1       louis    67
```

```
mydf[, 2]  # second column
```

```
[1] 67 85 32
```

```
mydf[1:2,]  # first two rows
```

```
 participant score
1       louis    67
2       paula    85
```

And these operations can be stacked, such as:

```
mydf[, 1][2] # first column, second entry
```

```
[1] paula
Levels: louis paula vincenzo
```

The last statement can be unpacked as follows: the first indexing operation extracts the first column. The output of this operation is itself a unidimensional vector, to which you can then apply another indexing operation.

What if you wanted to extract the row for the participant called Vincenzo? For this, logical statements can be used.

```
mydf[mydf$participant == 'vincenzo',]
```

```
  participant score
3 vincenzo       32
```

Let me paraphrase this command into plain English: 'Using the data frame mydf, extract only those rows for which the statement mydf$participant == 'vincenzo' returns a TRUE value.' Notice how the result of this indexing operation is a data frame with one row. Because the result is a data frame, you can use the dollar sign operator to further index a specific column, as in the following command:

```
mydf[mydf$participant == 'vincenzo',] $score
```

```
[1] 32
```

1.11. Loading in Files

When loading in files into your current working environment, R needs to know which folder on your computer to look at, what is called the 'working directory'. Use getwd() to check the current working directory.

```
getwd() # output specific to one's computer
```

```
[1] "/Users/bodo"
```

This is the folder on your machine where R currently 'looks at', and it is where it expects your files. You can look at the folder's content from within R, using the list.files() function. This should remind you of the ls() function. Whereas ls() displays the R-internal objects, list.files() displays R-external files.

```
list.files() # output not shown (specific to your machine)
```

To change your working directory to where your files are, you *can* use setwd(). Crucially, this command will be computer-specific, and it will differ between Mac/Linux and Windows. Rather than explaining all of this, I recommend you to set your working directory in RStudio, where you can find the menu item 'Set Working Directory' under the drop-down menu item 'Session'. Once you've clicked on 'Set Working Directory', navigate to the folder where the files for this book are (if you haven't downloaded those files yet, now is your chance!). It is OK if the files in the folder are not displayed or grayed out. You can still click 'Open' here as your goal right now is not to select the file, but the folder where the file is located at. Once the working directory has been set, you load the 'nettle_1999_climate.csv' file using `read.csv()` as follows. The .csv extension means that this is a comma-separated file (internally, columns are separated by commas). This dataset is taken from Nettle's (1999) book *Linguistic Diversity* and will be explained in more detail in later chapters.

```
nettle <- read.csv('nettle_1999_climate.csv')
```

If there's no error message, you have successfully loaded the file. If there *is* an error message, check whether you have typed the file name correctly. If that is the case, use `list.files()` to check whether the file is actually in the folder. If that is not the case, you may not have set the working directory successfully, which you can assess using `getwd()`.

Whenever you load a file into R, the next step should be to check its content. The `head()` function shows the first six rows of the `nettle` data frame (the 'head' of the data frame).

```
head(nettle)
```

```
    Country Population Area  MGS  Langs
1    Algeria      4.41 6.38 6.60     18
2     Angola      4.01 6.10 6.22     42
3  Australia      4.24 6.89 6.00    234
4 Bangladesh      5.07 5.16 7.40     37
5      Benin      3.69 5.05 7.14     52
6    Bolivia      3.88 6.04 6.92     38
```

The `tail()` function displays *last* six rows (the 'tail' of the data frame).

```
tail(nettle)
```

```
     Country Population Area  MGS  Langs
69 Venezuela      4.31 5.96 7.98     40
70   Vietnam      4.83 5.52 8.80     88
71     Yemen      4.09 5.72 0.00      6
72     Zaire      4.56 6.37 9.44    219
73    Zambia      3.94 5.88 5.43     38
74  Zimbabwe      4.00 5.59 5.29     18
```

It is important to discuss a few more points regarding file management for data analysis projects. First, when you quit RStudio, a 'Quit R session' question will pop up, asking you whether you want to save the content of your workspace. Click 'No'. Each time you open up R or RStudio, your new R session should open a new working environment and load in the required files. You don't want objects from previous analysis projects to be floating around, which may slow things down and cause naming conflicts (two objects having the same name). Instead, you want to keep all your data external to R.

There's also absolutely no problem if you override or messed up an R object within a session. Let's say you accidentally override the `Langs` column with NAs (missing values).

```
nettle$Langs <- NA
```

This is not a problem at all. Simply re-execute the entirety of your script up to the point where the mistake happened, and you will have everything back to where it was. For most simple analyses, there's not really any purpose for 'backing up' R objects.[7]

It is a good idea to structure your workflow around .csv files, as these are quite easy to deal with in R. Of course, R supports many other file types. For example, the 'example_file.txt' in the book's folder is a tab-separated file; that is, columns are separated by tabs (which are written '\t' computer-internally). You can use read.table() to load in this file as follows (ignore the warning message):[8]

```
mydf <- read.table('example_file.txt',
                   sep = '\t', header = TRUE)
```

```
Warning message:
In read.table("example_file.txt", sep = "\t", header = TRUE):
 incomplete final line found by readTableHeader on
'example_file.txt'
```

```
mydf
```

```
  amanda jeannette gerardo
1      3         1       2
2      4         5       6
```

The read.table() function requires you to specify a separator (the argument sep), which is '\t' for tabs in this case. The header = TRUE argument is required when the first row of the table contains column names. Sometimes, column name information is stored in a separate file, in which case you need header = FALSE.

There are too many file types to cover in this book. Often, Google will be your friend. Alternatively, you may resort to Excel to open up a file and save it in a format that you can easily read into R, such as .csv. However, let me tell you about a general strategy for dealing with file types where you don't know the internal structure. Load the file into R as text, using readLines().

```
x <- readLines('example_file.txt', n = 2)
```

```
x
```

```
[1] "amanda\tjeannette\tgerardo" "3\t1\t2"
```

---

7 The save() function allows you to save R objects in .RData files, as demonstrated here for the object mydf. You can use load() to load an .RData file into your current R session. Knowing about this comes in handy when you have to deal with computations that take a long time.
    save(mydf, file = 'mydataframe.RData')
8 Warning messages differ from error messages. A warning message happens when a command was executed but the function wants to warn you of something. An error message means that a command was aborted.

The n argument specifies the number of lines to be read. You often do not need more than two lines in order to understand the structure of an object. In this case, you can see that the first line contains column names, which tells you that header = TRUE is necessary. In addition, the second row contains tab delimiters '\t', which tells you that sep = '\t' is necessary. You can use this information to provide the right arguments to the read.table() function.

If you want to load in Excel files (.xls, .xlsx), there are various packages available. Other R packages exist for loading in SPSS or STATA files. However, all of these file types are proprietary, including .xls and .xlsx. This means that some company owns these file types. Whenever possible, try to avoid proprietary file types and use simple comma- or tab-separated files to manage your projects.

A note on how to use data in this book. For each chapter, new data will be loaded. It is perfectly fine to leave R open and work within one session for the entirety of this book. Alternatively, you can close R after working on a specific chapter (don't forget to *not* save your workspace). While it is OK for this book to work within one continued R session, you will always want to start a new R session in an actual data analysis, which helps to keep things neat.

1.12. Plotting

Let's begin by creating one of the most useful plots in all of statistics, the histogram. Figure 1.1 shows a histogram of the number of languages per country. The height of each rectangle (called 'bin') indicates the number of data points contained within the range covered by the rectangle, what is called the 'bin width'. In this case, there are

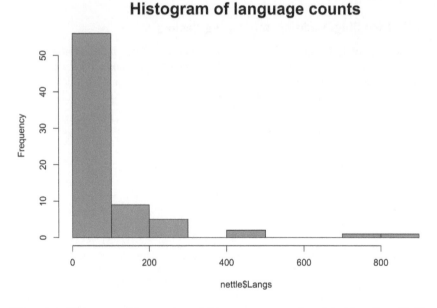

Figure 1.1. Histogram of the number of languages per country; data taken from Nettle (1999)

more than 50 countries that have between 0 and 100 languages. There are about 10 countries that have between 100 and 200 languages, and so on. There seems to be only a handful of countries with more than 700 languages.

The following command creates this plot, using the Langs column of the Nettle (1999) dataset.[9]

```
hist(nettle$Langs)
```

Let's rerun the histogram command and make the bins have a certain color, say, salmon.

```
hist(nettle$Langs, col = 'salmon')
```

The col argument is an optional argument of the hist() function. See what happens if you change the color to, say, 'steelblue'. If you wanted to have a look at what colors are pre-specified in R, type in colors() to the console. This is another function that doesn't require any arguments, similar to ls(), list.files(), or getwd(). The code below shows only the first six colors.

```
head(colors())
```

```
[1] "white"          "aliceblue"      "antiquewhite"
[4] "antiquewhite1"  "antiquewhite2"  "antiquewhite3"
```

These colors are the *named* colors of R. You can also specify hexadecimal color codes. Check what happens if you supply '#DD4433' to the col argument.

1.13. Installing, Loading, and Citing Packages

R is a community project with a massive amount of content made available by its active user base. New functions are assembled into libraries called 'packages', which can be installed using the install.packages() function. The following code installs the car package (Fox & Weisberg, 2011).

```
install.packages('car')
```

Once a package is installed,[10] you can load it like this:

```
library(car)
```

9 These language counts are a considerable abstraction, as it's not always clear whether a linguistic variety is best considered as a dialect of a language, or whether it should be counted as its own language.
10 There are many reasons why the installation of a package may fail. Installation problems can usually be fixed with the help of Google searches. A lot of the time, installation problems result from having an outdated R version. Alternatively, there may have been a new R version released recently, and the package developers are lagging behind in updating their package to the new version.

This makes the functions from the `car` package available in the current R session. If you close R and reopen it, you will have to reload the package. That is, packages are only loaded for a single session.

For reproducibility and to acknowledge the valuable work of the respective developers, it's important to cite each package and report its package version, which is demonstrated here for `car`:

```
citation('car')$textVersion
```

```
[1] "John Fox and Sanford Weisberg (2011). An {R} Compan-
ion to Applied Regression, Second Edition. Thousand Oaks
CA: Sage. URL: http://socserv.socsci.mcmaster.ca/jfox/
Books/Companion"
```

```
packageVersion('car')
```

```
[1] '3.0.0'
```

Speaking of citing, this is how you can retrieve the citation and version information for R itself. These should also be reported. The commands below provide abbreviated output thanks to the indexing statements '`$textVersion`' and '`$version.string`'. You can also simply run the `citation()` and `R.Version()` functions without these statements, which returns more extensive outputs.

```
citation()$textVersion
```

```
[1] "R Core Team (2018). R: A language and environment for
statistical computing. R Foundation for Statistical Com-
puting, Vienna, Austria. URL https://www.R-project.org/."
```

```
R.Version()$version.string
```

```
[1] "R version 3.5.0 (2018-04-23)"
```

1.14. Seeking Help

There's a help file for any R function. To access it, just put a question mark in front of a function's name, as demonstrated here for the `seq()` function, which is for generating number sequences.

```
?seq
```

I haven't introduced you to this function yet, but perhaps you can figure out how it works from the examples listed at the bottom of the help file (you will use this function in later chapters).

If you forgot a function name and wanted to search for it from within R, you can use the handy `apropos()` function. For example, running `apropos('test')` will display all functions that have the string `'test'` in their name. In the following, I only show you the first six of these:

```
head(apropos('test'))
```

```
[1] ".valueClassTest" "ansari.test"
[3] "bartlett.test"   "binom.test"
[5] "Box.test"        "chisq.test"
```

Often, copy-and-pasting warning or error messages from the console into Google will immediately direct you to a solution to any problems you may encounter. If that doesn't help, you can ask others for help, such as via stackoverflow.com. However, asking good questions is not easy and it's essential that you perform extensive Google searches before asking others.

Perhaps most importantly, when you encounter an error or warning message, you should never think that you are stupid. R is quite quirky and learning how to program isn't easy. If you encounter a problem, rest assured that there are many others who ran into the same problem. You should be aware of the fact that even very advanced R users constantly encounter error and warning messages. For example, the very experienced R programmers Wickham and Grolemund (2007: 7) write: "I have been writing R code for years, and every day I still write code that doesn't work!" This should show you that there's no reason to feel stupid when you run into a problem.

1.15. A Note on Keyboard Shortcuts

You are heavily advised to spend some time learning R/RStudio keyboard shortcuts. When wanting to be efficient with data, the mouse is your enemy! Your future self is going to thank you (yet again!) for the countless hours saved thanks to keyboard shortcuts. Here are some very handy shortcuts that I use frequently:

Shortcut	Action
Ctrl/Command + N	open new script
Ctrl/Command + Enter	run current line (send from script to console)
Alt/Option + Minus '-'	insert '<-'
Ctrl/Command + Alt/Option + I	insert code chunk (R markdown, see Chapter 2)
Ctrl/Command + Shift + M	insert pipe (tidyverse, see Chapter 2)

These shortcuts are specific to R/RStudio. On top of that, I hope that you already are accustomed to using general text editing shortcuts. If not, here are some very useful shortcuts:

Shortcut	Action
Shift + Left/Right	highlight subsection of text
Alt/Option + Left/Right (Mac)	move cursor by a word

Ctrl + Left/Right (Windows)	move cursor by a word
Command + Left/Right (Mac)	move cursor to beginning/end of line
Home/End (Windows & Linux)	move cursor to beginning/end of line
Ctrl + K	delete line from position of cursor onwards
Ctrl/Command + C	copy
Ctrl/Command + X	cut
Ctrl/Command + V	paste
Ctrl/Command + Z	undo
Ctrl/Command + A	select all

Incorporating these and other shortcuts into your workflow will save you valuable time and mental energy. In the long run, knowing lots of shortcuts will help you experience 'flow' when coding.

1.16. Your R Journey: The Road Ahead

I have just taught you the very basics of R. Think about learning R as learning a foreign language. Obviously, you cannot learn a whole language in a single chapter! And, of course, you're bound to forget things, which is OK! It's important that you persevere through your mistakes. When you get stuck, take a break and resume later. Let's talk about the most common problems you will encounter:

- You wrote something in your script, but you have forgotten to execute the command in R (that is, it hasn't been sent to the console yet).
- If you get an error message which says "object not found", you likely have mistyped the name of the object, or you forgot to execute some assignment command.
- If you get a "function not found" message, you either mistyped the function name, or the relevant package hasn't been loaded yet.
- Warning messages also frequently arise from not knowing what type of object you are working with: What's the object's dimensionality (rows, columns)? What type of vector are you dealing with (e.g., character, factor, numeric, logical)?
- Sometimes you may run into syntax issues, such as having forgotten a comma between arguments inside a function, or having forgotten a closing bracket.
- Many or most errors result from some sort of typo. Extensive use of the copy-and-paste shortcuts prevents many typos.

As a general rule of thumb, you should never believe R does as intended (Burns, 2011). It is good to develop a habit of constantly checking the objects in your working environment. You should ask yourself questions such as 'Did the function I just executed actually produce the intended result?' or 'What is actually contained in the object that I'm currently working with?'

From now on, it's just learning-by-doing and trial-and-error. I will teach you some more R in the next chapter. After that, the focus will be on the stats side of things. As you go along, many commands will be repeated again and again. The same way it helps you to repeat words when learning the vocab of a foreign language, R 'words' need to be used extensively before they sink in. And, also like learning a foreign language, there's no way around continuous practice. I'll help you along the way.

1.17. Exercises

1.17.1. Exercise 1: Familiarizing Yourself with Base Plotting

Type the following commands into a script and then execute them together:

```
plot(x = 1, y = 1, type = 'n',
     xlim = c(-2, 2), ylim = c(-2, 2))
points(x = -1, y = 1)
segments(x0 = -0.5, y0 = -1, x1 = 0.5, y1 = -1)
```

The first line opens up an empty plot with a point at the coordinates $x = 1$ and $y = 1$. The type = 'n' argument means that this point is not actually displayed. xlim and ylim specify the plot margins.

The points() function plots a single point at the specified coordinates. The segments() function plots a line segment. The beginning of the line segment is given by the arguments x0 and y0. The end points are given by x1 and y1.

What is displayed in your plotting window is actually a one-eyed smiley. By adding additional points() and segments(), can you create a more elaborate smiley? This exercise will help to wrap your head around working with coordinate systems.

1.17.2. Exercise 2: Swirl

The interactive swirl package teaches you R inside R.

```
install.packages('swirl')
library(swirl)
swirl()
```

Complete the first four modules of the R programming course. If you have extra time, complete the first five courses of the Exploratory Data Analysis course. Throughout your R journey, you can come back to swirl at any time and complete more courses.

1.17.3. Exercise 3: Spot-the-Error #1

Type the following two lines of code into your R script, exactly as they are printed here on the page. Then execute them. This will result in two error messages.

```
x_values <- c(1, 2 3, 4, 5, 6, 7, 8, 9)
mean_x <- mean(X_values)
```

Each line contains one error. Can you find them and correct them?[11]

11 I thank Márton Sóskuthy for this exercise idea.

1.17.4. Exercise 4: Spot-the-Error #2

Why does the following command return an NA value?

```
x <- c(2, 3, 4, '4')
mean(x)
```

Can you use the function as.numeric() to solve this problem?

1.17.5. Exercise 5: Spot-the-Error: #3

The following line of code tries to extract the row from the nettle data frame that contains information on the country Yemen. Why does this return an error and can you fix this?

```
nettle[nettle$Country = 'Yemen',  ]
```

1.17.6. Exercise 6: Indexing Data Frames

Gillespie and Lovelace (2017: 4) say that "R is notorious for allowing users to solve problems in many ways". In this section, you will learn a bunch of different ways of extracting information from the same data frame. Some of these ways are redundant, but knowing multiple paths to the same goal gives you flexibility in how to approach data analysis problems. This exercise also teaches you how indexing statements can be used recursively, stacked on top of each other.

```
head(nettle)   # display first 6 rows

     Country Population Area  MGS Langs
1    Algeria       4.41 6.38 6.60    18
2     Angola       4.01 6.10 6.22    42
3  Australia       4.24 6.89 6.00   234
4 Bangladesh       5.07 5.16 7.40    37
5      Benin       3.69 5.05 7.14    52
6    Bolivia       3.88 6.04 6.92    38
```

Next, the following statements extract information from this data frame. You haven't been taught all of these ways of indexing yet. However, try to understand what the corresponding code achieves and I'm sure you'll be able to figure it out.

Importantly, think about what is being extracted *first*, only *then* type in the command to see whether the output matches your expectations.

```
nettle[2, 5]
```

```
nettle[1:4,  ]

nettle[1:4, 1:2]

nettle[nettle$Country == 'Bangladesh',  ]

nettle[nettle$Country == 'Bangladesh', 5]

nettle[nettle$Country == 'Bangladesh',  ] [, 5]

nettle[nettle$Country == 'Bangladesh',  ] $Langs

nettle[nettle$Country == 'Bangladesh', 'Langs']

nettle[1:4,  ] $Langs[2]

nettle[1:4, c('Country', 'Langs')]

head(nettle[,])
```

2   The Tidyverse and Reproducible R Workflows

2.1. Introduction

This chapter serves two goals. First, I will introduce you to the tidyverse, a modern way of doing R. Second, I will introduce you to reproducible research practices. As part of this, I will talk about ways efficient workflows for analysis projects. A disclaimer is necessary: This chapter is very technical. If you feel overwhelmed, don't worry, as the concepts discussed here will come up again and again throughout the book.

The tidyverse is a modern way of doing R that is structured around a set of packages created by Hadley Wickham and colleagues. The idea is to facilitate interactive data analysis via functions that are more intuitive and 'tidier' than some of the corresponding base R functions. Let's begin by installing and loading in the `tidyverse` package (Wickham, 2017).

```
install.packages('tidyverse')

library(tidyverse)
```

The tidyverse package is actually just a handy way of installing and loading lots of packages at the same time; namely, all those packages that are part of the tidyverse.[1] I will walk you through some, but not all, of the core members and some of their most useful functions. In this chapter, you will learn about `tibble` (Müller & Wickham, 2018), `readr` (Wickham, Hester, & François, 2017), `dplyr` (Wickham, François, Henry, & Müller, 2018), `magrittr` (Milton Bache & Wickham, 2014), and `ggplot2` (Wickham, 2016).

---

[1] You could also load the individual packages from the tidyverse separately. These are the tidyverse packages discussed in this chapter:
```
library(tibble)
library(readr)
library(dplyr)
library(magrittr)
library(ggplot2)
```

2.2. `tibble` and `readr`

Tibbles from the `tibble` package (Müller & Wickham, 2018) are a modern take on data frames. They are like base R data frames, but better. Specifically, they confer the following four advantages:

- For text, tibbles default to character vectors rather than factor vectors, which is useful because character vectors are easier to manipulate.
- When typing the name of a tibble into the console, only the first ten rows are displayed, saving you a lot of `head()` function calls.
- Tibbles additionally display row and column numbers, saving you a lot of `nrow()` and `ncol()` function calls.
- Finally, displaying a tibble also reveals how each column in a tibble is coded (character vector, numeric vector, etc.).

All of these appear to be minor cosmetic adjustments. Together, however, these small changes end up saving you lots of time (and typing!). To see tibbles in action, let's load in a data frame and convert it to a tibble with `as_tibble()`:

```
# Load data:

nettle <- read.csv('nettle_1999_climate.csv')

# Convert data frame to tibble:

nettle <- as_tibble(nettle)
```

Type the name of the tibble:

```
nettle
```

```
# A tibble: 74 x 5
   Country     Population  Area   MGS  Langs
   <fct>            <dbl> <dbl> <dbl>  <int>
 1 Algeria           4.41  6.38  6.6     18
 2 Angola            4.01  6.1   6.22    42
 3 Australia         4.24  6.89  6      234
 4 Bangladesh        5.07  5.16  7.4     37
 5 Benin             3.69  5.05  7.14    52
 6 Bolivia           3.88  6.04  6.92    38
 7 Botswana          3.13  5.76  4.6     27
 8 Brazil            5.19  6.93  9.71   209
 9 Burkina Faso      3.97  5.44  5.17    75
10 CAR               3.5   5.79  8.08    94
# ... with 64 more rows
```

Notice that, besides printing out only ten rows, the number of rows (74) and columns (5) is stated, as well as information about vector classes. The `<dbl>` stands for 'double', which is computer-science-speak for a particular type of numeric vector—just treat doubles as numeric vectors. The <int> stands for 'integer'; <fct> stands for 'factor'. But wait, didn't I just tell you that tibbles default to character vectors? Why is the Country column coded as a factor? The culprit here is the base R function read.csv(), which automatically interprets any text column as factor. So, before the data frame was converted into a tibble, the character-to-factor conversion has already happened.

To avoid this and save yourself the conversion step, use read_csv() from the readr package (Wickham et al., 2017) (notice the underscore in the function name).

```
nettle <- read_csv('nettle_1999_climate.csv')
```

```
Parsed with column specification:
cols(
  Country = col_character(),
  Population = col_double(),
  Area = col_double(),
  MGS = col_double(),
  Langs = col_integer()
)
```

The read_csv() function tells you how columns have been 'parsed', that is, how particular columns from the file were converted to particular vector types. In addition, read_csv() creates tibbles by default. Let's verify this:

```
nettle
```

```
# A tibble: 74 x 5
     Country     Population Area    MGS Langs
     <chr>            <dbl> <dbl> <dbl> <int>
 1 Algeria           4.41  6.38  6.6     18
 2 Angola            4.01  6.1   6.22    42
 3 Australia         4.24  6.89  6      234
 4 Bangladesh        5.07  5.16  7.4     37
 5 Benin             3.69  5.05  7.14    52
 6 Bolivia           3.88  6.04  6.92    38
 7 Botswana          3.13  5.76  4.6     27
 8 Brazil            5.19  6.93  9.71   209
 9 Burkina Faso      3.97  5.44  5.17    75
10 CAR               3.5   5.79  8.08    94
# ... with 64 more rows
```

Notice how the Country column is now coded as a character vector, rather than as a factor vector.

In addition, read_csv() runs faster than read.csv(), and it provides a progress bar for large datasets. For files that are not comma-separated files (.csv), use read_delim(), for which the delim argument specifies the type of separator. The following command loads the tab-delimited 'example_file.txt':

```
x <- read_delim('example_file.txt', delim = '\t')
```

```
Parsed with column specification:
cols(
  amanda = col_integer(),
  jeannette = col_integer(),
  gerardo = col_integer()
)
```

```
x
```

```
# A tibble: 2 x 3
  amanda jeannette gerardo
   <int>     <int>   <int>
1      3         1       2
2      4         5       6
```

2.3. dplyr

The dplyr package (Wickham et al., 2018) is the tidyverse's workhorse for changing tibbles. The filter() function filters rows. For example, the following command reduces the nettle tibble to only those rows with countries that have more than 500 languages.

```
filter(nettle, Langs > 500)
```

```
# A tibble: 2 x 5
  Country          Population  Area   MGS Langs
  <chr>                 <dbl> <dbl> <dbl> <int>
1 Indonesia              5.27  6.28  10.7   701
2 Papua New Guinea       3.58  5.67  10.9   862
```

Alternatively, you may be interested in the data for a specific country, such as Nepal:

```
filter(nettle, Country == 'Nepal')
```

```
# A tibble: 1 x 5
  Country Population  Area   MGS Langs
  <chr>        <dbl> <dbl> <dbl> <int>
1 Nepal         4.29  5.15  6.39   102
```

So, the filter() function takes the input tibble as its first argument. The second argument is a logical statement that you use to put conditions on the tibble, thus restricting the data to a subset of rows.

The select() function is used to select columns. Just list all the columns you want to select, separated by commas. Notice that the original column order does not need to be obeyed, which means that select() can also be used to reorder tibbles.

```
select(nettle, Langs, Country)
```

```
# A tibble: 74 x 2
   Langs Country
   <int> <chr>
1     18 Algeria
2     42 Angola
3    234 Australia
4     37 Bangladesh
5     52 Benin
6     38 Bolivia
7     27 Botswana
8    209 Brazil
9     75 Burkina Faso
10    94 CAR
# ... with 64 more rows
```

Using the minus sign in front of a column name excludes that column.

```
select(nettle, -Country)
```

```
# A tibble: 74 x 4
  Population  Area   MGS Langs
       <dbl> <dbl> <dbl> <int>
1       4.41  6.38  6.6     18
2       4.01  6.1   6.22    42
3       4.24  6.89  6      234
4       5.07  5.16  7.4     37
5       3.69  5.05  7.14    52
6       3.88  6.04  6.92    38
7       3.13  5.76  4.6     27
8       5.19  6.93  9.71   209
9       3.97  5.44  5.17    75
10      3.5   5.79  8.08    94
# ... with 64 more rows
```

Use the colon operator to select consecutive columns, such as all the columns from Area to Langs.

```
select(nettle, Area:Langs)
```

```
# A tibble: 74 x 3
   Area   MGS Langs
  <dbl> <dbl> <int>
1  6.38  6.6     18
2  6.1   6.22    42
3  6.89  6      234
```

```
 4   5.16  7.4      37
 5   5.05  7.14     52
 6   6.04  6.92     38
 7   5.76  4.6      27
 8   6.93  9.71    209
 9   5.44  5.17     75
10   5.79  8.08     94
# ... with 64 more rows
```

To summarize the two dplyr functions introduced so far: filter() is used to filter rows; select() is used to select columns.

The rename() function can be used to change the name of existing columns. Each argument is structured as follows: 'New column name equals old column name.' For example, the following code shortens the name of the Population column to Pop.

```
nettle <- rename(nettle, Pop = Population)

nettle
```

```
# A tibble: 74 x 5
   Country         Pop   Area  MGS   Langs
   <chr>         <dbl> <dbl> <dbl>   <int>
 1 Algeria        4.41  6.38  6.6      18
 2 Angola         4.01  6.1   6.22     42
 3 Australia      4.24  6.89  6       234
 4 Bangladesh     5.07  5.16  7.4      37
 5 Benin          3.69  5.05  7.14     52
 6 Bolivia        3.88  6.04  6.92     38
 7 Botswana       3.13  5.76  4.6      27
 8 Brazil         5.19  6.93  9.71    209
 9 Burkina Faso   3.97  5.44  5.17     75
10 CAR            3.5   5.79  8.08     94
# ... with 64 more rows
```

The mutate() function can be used to change the content of a tibble. For example, the following command creates a new column Lang100, which is specified to be the Langs column divided by 100.

```
nettle <- mutate(nettle, Lang100 = Langs / 100)

nettle
```

```
# A tibble: 74 x 6
   Country    Population  Area   MGS Langs Lang100
   chr>            <dbl> <dbl> <dbl> <int>   <dbl>
 1 Algeria          4.41  6.38  6.6     18    0.18
 2 Angola           4.01  6.1   6.22    42    0.42
```

```
 3 Australia       4.24  6.89  6       234  2.34
 4 Bangladesh      5.07  5.16  7.4      37  0.37
 5 Benin           3.69  5.05  7.14     52  0.52
 6 Bolivia         3.88  6.04  6.92     38  0.38
 7 Botswana        3.13  5.76  4.6      27  0.27
 8 Brazil          5.19  6.93  9.71    209  2.09
 9 Burkina Faso    3.97  5.44  5.17     75  0.75
10 CAR             3.5   5.79  8.08     94  0.94
# ... with 64 more rows
```

Finally, arrange() can be used to order a tibble in ascending or descending order. Let's use this function to look at the countries with the largest and the smallest number of languages.

```
arrange(nettle, Langs)   # ascending
```

```
# A tibble: 74 x 6
   Country       Population  Area   MGS Langs Lang100
   <chr>              <dbl> <dbl> <dbl> <int>   <dbl>
 1 Cuba                4.03  5.04  7.46     1    0.01
 2 Madagascar          4.06  5.77  7.33     4    0.04
 3 Yemen               4.09  5.72  0        6    0.06
 4 Nicaragua           3.6   5.11  8.13     7    0.07
 5 Sri Lanka           4.24  4.82  9.59     7    0.07
 6 Mauritania          3.31  6.01  0.75     8    0.08
 7 Oman                3.19  5.33  0        8    0.08
 8 Saudi Arabia        4.17  6.33  0.4      8    0.08
 9 Honduras            3.72  5.05  8.54     9    0.09
10 UAE                 3.21  4.92  0.83     9    0.09
# ... with 64 more rows
```

```
arrange(nettle, desc(Langs))   # descending
```

```
# A tibble: 74 x 6
   Country          Population  Area   MGS Langs Lang100
   <chr>                 <dbl> <dbl> <dbl> <int>   <dbl>
 1 Papua New Guinea       3.58  5.67  10.9   862    8.62
 2 Indonesia              5.27  6.28  10.7   701    7.01
 3 Nigeria                5.05  5.97  7      427    4.27
 4 India                  5.93  6.52  5.32   405    4.05
 5 Cameroon               4.09  5.68  9.17   275    2.75
 6 Mexico                 4.94  6.29  5.84   243    2.43
 7 Australia              4.24  6.89  6      234    2.34
 8 Zaire                  4.56  6.37  9.44   219    2.19
 9 Brazil                 5.19  6.93  9.71   209    2.09
10 Philippines            4.8   5.48  10.3   168    1.68
# ... with 64 more rows
```

2.4. ggplot2

The ggplot2 package (Wickham, 2016) is many people's favorite package for plotting. The logic of ggplot2 takes some time to get used to, but once it clicks you'll be able to produce beautiful plots very quickly.

Let's use ggplot2 to graphically explore the relation between climate and linguistic diversity. Nettle (1999) discusses the intriguing idea that linguistic diversity is correlated with climate factors. The proposal is that countries with lower ecological risk have more different languages than countries with higher ecological risk. A subsistence farmer in the highlands of Papua New Guinea lives in a really fertile environment where crops can be grown almost the entire year, which means that there is little reason to travel. When speakers stay local and only speak with their immediate neighbors, their languages can accumulate differences over time which would otherwise be levelled through contact.

Nettle (1999) measured ecological risk by virtue of a country's 'mean growing season' (listed in the MGS column), which specifies how many months per year one can grow crops.

Let's plot the number of languages (Langs) against the mean growing season (MGS). Type in the following command and observe the result, which is shown in Figure 2.1 (left plot)—a detailed explanation will follow.

```
ggplot(nettle) +
  geom_point(mapping = aes(x = MGS, y = Langs))
```

Like most other functions from the tidyverse, the ggplot() function takes a tibble as its first argument. However, the rest of how this function works takes some time to wrap your head around. In particular, you have to think about your plot in a different way, as a layered object, where the data is the substrate, with different visual representations (shapes, colors, etc.) layered on top.

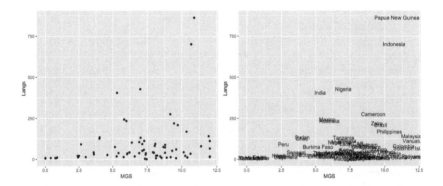

Figure 2.1. Left: scatterplot of the number of languages per mean growing season; right: the same scatterplot but with text; each data point is represented by the respective country name

In this case, the data are `MGS` and `Langs` taken from the `nettle` tibble. However, the same data could be displayed in any number of different ways. How exactly the data is visualized is specified by what is called a 'geom', a geometric object. Each of the plots that you commonly encounter in research papers (histograms, scatterplots, bar plots, etc.) has their own geom, that is, their own basic shape. In the case of scatterplots, for example, the data is mapped to points in a 2D plane (as seen in Figure 2.1). In the case of bar plots (see below), the data is mapped to the height of the bars. As you proceed in this book, you will learn about other geoms, such as `geom_boxplot` (the data is mapped to boxes) or `geom_text` (the data is mapped to text).

So, the geom indicates the primary shape which is used to visually represent the data from a tibble. The code above adds a point geom to the plot via the `geom_point()` function. For plotting points in a two-dimensional plane, one needs both x and y values. The function `aes()` specifies the 'aesthetic mappings', which characterize which aspect of the data is mapped onto which aspect of the geom. In this case, `MGS` is mapped onto the x-values, and `Langs` is mapped onto the y-values. Think about it this way: the geometric objects are flying on top of the data substrate, but they need to know which specific aspects of the data to draw from, and this is specified by the aesthetic mappings.

Now, I recognize that the logic of the `ggplot2` package may appear confusing at this stage. I was definitely confused when I encountered `ggplot2` for the first time. As you proceed and type in more different plotting commands, you will slowly get the hang of things.

Let's learn a bit more about the aesthetic mappings. It turns out that you can either specify them inside a geom, or inside the main `ggplot()` function call. The latter allows multiple geoms to draw from the same mappings, which will become useful later. Notice, furthermore, that I omitted the argument name 'mapping'.

```
ggplot(nettle, aes(x = MGS, y = Langs)) +
  geom_point()
```

Finally, let's create a plot where instead of points, the country names are displayed at their respective x, y coordinates. Let's see what happens when `geom_point()` is replaced with `geom_text()`.

```
ggplot(nettle, aes(x = MGS, y = Langs)) +
  geom_text()
```

Error: geom_text requires the following missing aesthetics: label

The problem here is that `geom_text()` needs an additional aesthetic mapping. In particular, it needs to know which column is mapped to the actual text (`label`) shown in the plot, which is given in the `Country` column.

```
ggplot(nettle, aes(x = MGS, y = Langs, label = Country)) +
  geom_text()
```

The result is shown in the right plot of Figure 2.1. To save a plot, use `ggsave()` after a `ggplot2` command. For example, the following command saves the text plot into the file `nettle.png` with the width:height ratio of 8:6 (measurement units are given in inches by default).

```
ggsave('nettle.png', width = 8, height = 6)
```

To create the two-plot arrangement displayed in Figure 2.1, use the `gridExtra` package (Auguie, 2017).

```
# Create plots and save them in plot1 and plot2:

plot1 <- ggplot(nettle) +
    geom_point(mapping = aes(x = MGS, y = Langs))

plot2 <- ggplot(nettle,
          aes(x = MGS, y = Langs, label = Country)) +
    geom_text()

# Plot double plot:

library(gridExtra)
grid.arrange(plot1, plot2, ncol = 2)
```

The respective plots are saved into two objects, `plot1` and `plot2`, which are then used as arguments of the `grid.arrange()` function. The additional argument `ncol = 2` specifies a plot arrangement with two columns.

2.5. Piping with `magrittr`

A final component of the tidyverse relevant to us is the 'pipe', which is represented by the symbol sequence '`%>%`'. This functionality is unlocked by the tidyverse package `magrittr` (Milton Bache & Wickham, 2014).

Imagine a conveyor belt where the output of one function serves as the input to another function. The following code chunk exemplifies such a pipeline. The tibble `nettle` is first piped to the `filter()` function, which reduces the tibble to only those countries where one can grow crops for more than eight months of the year (MGS > 8). The filtered tibble is then piped to `ggplot()`, which results in a truncated version of Figure 2.1.

```
# Plotting pipeline with %>%:

nettle %>%
  filter(MGS > 8) %>%
  ggplot(aes(x = MGS, y = Langs, label = Country)) +
    geom_text()
```

Notice that the tibble containing the data only had to be mentioned once at the beginning of the pipeline, which saves a lot of typing. The true benefits of pipelines will be felt more strongly later on in this book.

2.6. A More Extensive Example: Iconicity and the Senses

This section guides you through the first steps of an analysis that was published in Winter, Perlman, Perry, and Lupyan (2017). This study investigated iconicity, the resemblance between a sign's form and its meaning. For example, the words *squealing*, *banging*, and *beeping* resemble the sounds they represent (also known as onomatopoeia, a specific form of iconicity). It has been proposed that sound concepts are more expressible via iconic means than concepts related to the other senses, such as sight, touch, smell, or taste. This may be because auditory ideas are easier to express via imitation in an auditory medium, speech.

To test this idea, we used sensory modality ratings from Lynott and Connell (2009), paired with our own set of iconicity ratings (Perry, Perlman, Winter, Massaro, & Lupyan, 2017; Winter et al., 2017). Let's load in the respective datasets and have a look at them.

```
icon <- read_csv('perry_winter_2017_iconicity.csv')
mod <- read_csv('lynott_connell_2009_modality.csv')
```

Let's check the content of both files, starting with the icon tibble. Depending on how wide your console is, more or fewer columns will be shown. Also, some numbers may be displayed differently. For example, the raw frequency 1041179 of the article *a* could be displayed in the abbreviated form 1.04e6 (this notation will be explained in more detail in Chapter 11, fn. 1).

```
icon
```

```
# A tibble: 3,001 x 8
     Word     POS          SER CorteseImag  Conc  Syst     Freq
     <chr>    <chr>      <dbl>       <dbl> <dbl> <dbl>    <int>
 1 a        Grammati... NA               NA  1.46    NA 1.04e6
 2 abide    Verb        NA               NA  1.68    NA 1.38e2
 3 able     Adjective  1.73             NA  2.38    NA 8.15e3
 4 about    Grammati... 1.2              NA  1.77    NA 1.85e5
 5 above    Grammati... 2.91             NA  3.33    NA 2.49e3
 6 abrasive Adjective  NA               NA  3.03    NA 2.30e1
 7 absorbe... Adjective NA               NA  3.1     NA 8.00e0
 8 academy  Noun        NA               NA  4.29    NA 6.33e2
 9 accident Noun        NA               NA  3.26    NA 4.15e3
10 accordi... Noun       NA               NA  4.86    NA 6.70e1
# ... with 2,991 more rows, and 1 more variable:
#   Iconicity <dbl>
```

The only three relevant columns for now are `Word`, `POS`, and `Iconicity`. Let's reduce the tibble to these columns using `select()`.

```
icon <- select(icon, Word, POS, Iconicity)

icon
```

```
# A tibble: 3,001 x 3
    Word        POS             Iconicity
    <chr>       <chr>               <dbl>
 1 a           Grammatical         0.462
 2 abide       Verb                0.25
 3 able        Adjective           0.467
 4 about       Grammatical        -0.1
 5 above       Grammatical         1.06
 6 abrasive    Adjective           1.31
 7 absorbent   Adjective           0.923
 8 academy     Noun                0.692
 9 accident    Noun                1.36
10 accordion   Noun               -0.455
# ... with 2,991 more rows
```

The dataset contains 3,001 words and their iconicity ratings. The `POS` column contains part-of-speech tags which were generated based on the SUBTLEX subtitle corpus of English (Brysbaert, New, & Keuleers, 2012). What about the content of the `Iconicity` column? In our rating task, we asked participants to rate words on a scale from −5 ('the word sounds like the opposite of what it means') to +5 ('the word sounds exactly like what it means'). The iconicity value of each word is an average of the ratings of multiple native speakers. Let's have a look at the range of this variable.

```
range(icon$Iconicity)
```

```
[1] -2.800000 4.466667
```

So, the lowest iconicity score is −2.8, the largest is +4.5 (rounded). This perhaps suggests that the iconicity ratings are skewed towards the positive end of the scale. To get a more complete picture, let's draw a histogram (depicted in Figure 2.2). Execute the following `ggplot2` code snippet—an explanation will follow.

```
ggplot(icon, aes(x = Iconicity)) +
  geom_histogram(fill = 'peachpuff3') +
  geom_vline(aes(xintercept = 0), linetype = 2) +
  theme_minimal()
```

```
'stat_bin()' using 'bins = 30'. Pick better value with
'binwidth'.
```

The warning message on binwidth can safely be ignored in this case. The code pipes the `icon` tibble to `ggplot()`. To draw a histogram, you just need only one aesthetic, namely, an aesthetic that maps the data to the relevant *x*-values. The `fill`

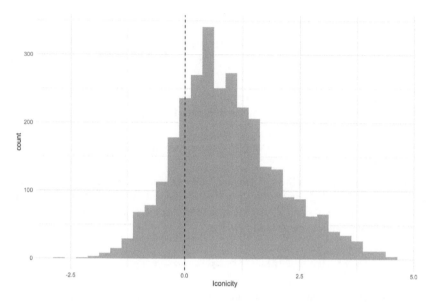

Figure 2.2. Histogram of iconicity values from Winter et al. (2017) and Perry et al. (2017)

of the histogram is specified to have the color `'peachpuff3'`. `geom_vline()` plots a vertical line at 0, which is specified by the `xintercept`. The optional argument `linetype = 2` makes the line dashed. Finally, adding `theme_minimal()` to the plot makes everything look nice. Themes are high-level plotting commands that change various aspects of the entire plot, such as the background color (notice the gray tiles have become white) and the grid lines. To see the effects of themes, explore what happens if you rerun the plot with `theme_linedraw()` or `theme_light()` instead. Themes save you a lot of typing, because you can avoid having to set all graphical parameters individually.

Next, let's check the `mod` tibble, which contains the modality norms from Lynott and Connell (2009).

```
mod
```

```
# A tibble: 423 x 9
   PropertyBritish Word       DominantModality Sight Touch
   <chr>           <chr>      <chr>            <dbl> <dbl>
 1 abrasive        abrasive   Haptic            2.89  3.68
 2 absorbent       absorbent  Visual            4.14  3.14
 3 aching          aching     Haptic            2.05  3.67
 4 acidic          acidic     Gustatory         2.19  1.14
 5 acrid           acrid      Olfactory         1.12  0.625
 6 adhesive        adhesive   Haptic            3.67  4.33
 7 alcoholic       alcoholic  Gustatory         2.85  0.35
```

```
 8 alive            alive    Visual              4.57 3.81
 9 amber            amber    Visual              4.48 0.524
10 angular          angular  Visual              4.29 4.10
# ... with 413 more rows, and 4 more variables:
#   Sound <dbl>, Taste <dbl>, Smell <dbl>,
#   ModalityExclusivity <dbl>
```

One feature of tibbles is that they only show as many columns as fit into your console. To display all columns, type the following (the extended output is not shown in the book):

```
mod %>% print(width = Inf)   # output not shown in book
```

The width argument allows you to control how many columns are displayed (thus, expanding or shrinking the 'width' of a tibble). By setting the width to the special value Inf (infinity), you display as many columns as there are in your tibble.

You may also want to display all rows:

```
mod %>% print(n = Inf)   # output not shown
```

The argument for rows is called n in line with the statistical convention to use the letter '*N*' to represent the number of data points.

For the present purposes, only the Word and DominantModality columns are relevant to us. The columns labeled Sight, Touch, Sound, Taste, and Smell contain the respective ratings for each sensory modality (on a scale from 0 to 5). Just as was the case with the iconicity rating study, these are averages over the ratings from several different native speakers. The content of the DominantModality column is determined by these ratings: for example, the first word, *abrasive*, is categorized as 'haptic' because its touch rating (3.68) is higher than the ratings for any of the other sensory modalities. Again, let's reduce the number of columns with select().[2]

```
mod <- select(mod, Word, DominantModality:Smell)

mod
```

```
# A tibble: 423 x 7
    Word        Dominant      Sight Touch Sound  Taste   Smell
                Modality
    <chr>       <chr>         <dbl> <dbl> <dbl>  <dbl>   <dbl>
 1  abrasive    Haptic        2.89  3.68  1.68   0.579   0.579
 2  absorbent   Visual        4.14  3.14  0.714  0.476   0.476
 3  aching      Haptic        2.05  3.67  0.667  0.0476  0.0952
 4  acidic      Gustatory     2.19  1.14  0.476  4.19    2.90
 5  acrid       Olfactory     1.12  0.625 0.375  3       3.5
```

---

2 It's usually a good idea to spend considerable time getting the data in shape. The more time you spend preparing your data, the less trouble you will have later.

```
 6 adhesive   Haptic     3.67  4.33   1.19  0.905  1.76
 7 alcoholic  Gustatory  2.85  0.35   0.75  4.35   4.3
 8 alive      Visual     4.57  3.81   4.10  1.57   2.38
 9 amber      Visual     4.48  0.524  0.143 0.571  0.857
10 angular    Visual     4.29  4.10   0.25  0.0476 0.0476
# ... with 413 more rows
```

To save yourself some typing further down the line, use `rename()` to shorten the name of the `DominantModality` column.

```
mod <- rename(mod, Modality = DominantModality)
```

Of course, you *could* have just changed the name in the spreadsheet before loading it into R. However, many times you work with large data files that are generated by some piece of software (such as E-Prime for psycholinguistic experiments, or Qualtrics for web surveys), in which case you want to leave the raw data untouched. It is usually a good idea to perform as much of the data manipulation as possible from within R. Even something as simple as renaming a column is a manipulation of the data, and it should be recorded in your scripts for the sake of reproducibility (more on this topic later).

When engaging with new datasets, it's usually a good idea to spend considerable time familiarizing yourself with their contents. For bigger datasets, it makes sense to check some random rows with the `dplyr` function `sample_n()`.

```
sample_n(mod, 4)   # shows 4 random rows, use repeatedly
```

```
# A tibble: 4 x 7
   Word        Modality  Sight Touch Sound Taste Smell
   <chr>       <chr>     <dbl> <dbl> <dbl> <dbl> <dbl>
1 thumping     Auditory  2.62  2.52  3.90  0.143 0.190
2 transparent  Visual    4.81  0.619 0.25  0.143 0.143
3 empty        Visual    4.75  3.6   1.65  0.25  0.15
4 spicy        Gustato.. 1.67  0.429 0.333 5     4.24
```

To assess whether the sensory modalities differ in iconicity, it is necessary to merge the two tibbles. The `left_join()` function call below takes two tibbles as argument, 'joining' the second tibble ('to the right') into the first tibble ('to the left').

```
both <- left_join(icon, mod)
```

```
Joining, by = "Word"
```

The `left_join()` function is smart and sees that both tibbles contain a `Word` column, which is then used for matching the respective data. If the identifiers for matching have different names, you can either rename the columns so that they match, or you can use the by argument (check the `?left_join()` help file).

Next, let's filter the dataset so that it only includes adjectives, verbs, and nouns (there are very few adverbs and grammatical words, anyway). For this, the very useful

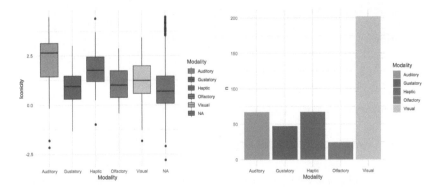

Figure 2.3. Left: Boxplot of iconicity ratings by sensory modality; right: Bar plot of word counts showing the over-representation of visual words in English

%in% function comes in handy. The function needs two vectors to compare, and it then checks whether the elements of the second vector are matched in the first vector.[3]

```
both <- filter(both,
          POS %in% c('Adjective', 'Verb', 'Noun'))
```

To put this command into plain English, one might paraphrase it as follows: 'Of the both tibble, filter only those rows for which the content of the POS column is in the set of adjectives, verbs, and nouns.'

Our main research question is whether iconicity ratings differ by modality. One option is to visualize this relationship with a boxplot, which will be explained in more detail in Chapter 3. A boxplot shows distributions as a function of categorical variable on the *x*-axis, as shown in Figure 2.3 (left plot). The following command maps Modality to the *x*-axis and Iconicity to the *y*-axis. Finally, an additional aesthetic maps the Modality categories onto the fill argument, which assigns each sensory modality a unique color.

```
ggplot(both,
     aes(x = Modality, y = Iconicity, fill = Modality)) +
     geom_boxplot() + theme_minimal()
```

The boxplot shows that the bulk of 'auditory' words (sound-related) have higher iconicity ratings than words for the other senses. 'Haptic' words (touch-related) also

---

3 To understand the set function %in%, check out what the following commands do:
```
c('A', 'B', 'C') %in% c('A', 'A', 'C', 'C')
```
... and then the reverse:
```
c('A', 'A', 'C', 'C') %in% c('A', 'B', 'C')
```
Basically, the %in% function is necessary when you want to use '==', but there are two things on the right-hand side of the equation. For the time being, it's OK to think of %in% as a generalized '=='.

have high iconicity ratings, and so on (for a discussion of these results, see Winter et al., 2017).

The right-most box displays the distribution of iconicity ratings for those words that couldn't be matched to a sensory modality because the word wasn't rated in Lynott and Connell's (2009) study. Let's exclude these missing values using `filter()`.[4] The `is.na()` function inside the following `filter()` command returns `TRUE` for any data point that is missing, and `FALSE` for any data point that is complete. The exclamation sign '`!`' inverts these logical values so that you only get the complete cases, that is, those cases that are *not* NA.[5] The code below also uses piping. The `both` tibble is piped to `filter()`, the output of which is then piped to `ggplot()`.

```
both %>% filter(!is.na(Modality)) %>%
  ggplot(aes(x = Modality, y = Iconicity,
             fill = Modality)) +
  geom_boxplot() + theme_minimal()
```

Let's explore another aspect of this data, which is discussed in more detail in Lievers and Winter (2017) and Winter, Perlman, and Majid (2018). In particular, it has been proposed that some sensory modalities are easier to talk about than others (Levinson & Majid, 2014), and that the English language makes more visual words available to its speakers than words for the other sensory modalities (see also Majid & Burenhult, 2014). To investigate this feature of the English language, have a look at the counts of words per sensory modality with `count()`.

```
both %>% count(Modality)
```

```
# A tibble: 6 x 2
  Modality       n
  <chr>      <int>
1 Auditory      67
2 Gustatory     47
3 Haptic        67
4 Olfactory     24
5 Visual       202
6 <NA>        2389
```

Ignoring the NAs for the time being, this tibble shows that there are overall more visual words. This was already noted by Lynott and Connell (2009).

Let's make a bar plot of these counts. As before, the `filter()` function is used to exclude NAs. The geom for bar plots is `geom_bar()`. You need to specify the additional argument `stat = 'identity'`. This is because the `geom_bar()` function

---

4 Generally, you should be concerned about missing values. In particular, you always need to ask *why* certain values are missing.

5 Alternatively, you could use the function `complete.cases()`, which returns `TRUE` for complete cases and `FALSE` for missing values. In that case, you wouldn't have to use the negation operation '`!`'. That said, I prefer `!is.na()` because it takes less time to type than `complete.cases()` ...

likes to perform its own statistical computations by default (such as counting). With the argument `stat = 'identity'`, you instruct the function to treat the data as is. Figure 2.4 (right plot) shows the resulting bar plot.

```
both %>% count(Modality) %>%
  filter(!is.na(Modality)) %>%
  ggplot(aes(x = Modality, y = n, fill = Modality)) +
  geom_bar(stat = 'identity') + theme_minimal()
```

To exemplify why you had to specify `stat = 'identity'` in the last code chunk, have a look at the following code, which reproduces Figure 2.4 in a different way.

```
both %>% filter(!is.na(Modality)) %>%
  ggplot(aes(Modality, fill = Modality)) +
  geom_bar(stat = 'count') + theme_minimal()
```

In this case, the `geom_bar()` function does the counting. This pipeline is a bit more concise since you don't have to create an intervening table of counts.

2.7. R Markdown

Chapter 1 introduced you to .R script files. Now you will learn about R markdown files, which have the extension .Rmd. R markdown files have more expressive power than regular R scripts, and they facilitate the reproducibility of your code. Basically, an R markdown file allows you to write plain text alongside R code. Moreover, you can 'knit' your analysis easily into a html file, which is done via the `knitr` package (Xie, 2015, 2018). The resulting html file contains your text, R code, and output. It is extremely useful to have all of this in one place. For example, when you write up your results, it'll be much easier to read off the numbers from a knitted markdown report, rather than having to constantly re-create the output in the console, which is prone to error. Moreover, the report can easily be shared with others. This is useful when collaborating—if your collaborators also know R, they can not only see the results, but they'll also know how you achieved these results. R markdown is a standard format for sharing your analysis on publicly accessible repositories, such as GitHub and the Open Science Framework.

To create an R markdown file in RStudio, click 'File', 'New File', and then 'R Markdown …' For now, simply leave the template as it is. See what happens if you press the 'knit' button (you will be asked to save the file first). This will create an html report of your analysis that shows the code, as well as the markdown content.

Let me guide you through some R markdown functionality. First, you can write plain text the same way you would do in a text editor. Use this text to describe your analysis. In addition, there are code chunks, which always begin with three ' ' ' (backward ticks, or the grave accent symbol). The R code goes inside the chunk. Any R code outside of code chunks won't be executed when knitting the file.

```
'''{r}
# R code goes in here
'''
```

You can specify additional options for each code chunk. For example, the code chunk below will print results, but not messages (such as warning messages, or package loading messages), and it will also 'cache' the results of running the code chunk. Caching means that the result of all computations conducted in this code chunk will be saved to memory files outside of the markdown script. This way, the computations don't have to be re-executed the next time you knit the markdown file. The argument cache = TRUE is useful when a particular code chunk takes a lot of time to run.

```
'''{r message = FALSE, cache = TRUE}
# R code goes in here
'''
```

The following code chunk, named myplot, does not print the R code (echo = FALSE). The additional code chunk options specify the width and height of the plot that will be displayed in the knitted html file.

```
'''{r echo = FALSE, fig.width = 8, fig.height = 6}
# plotting commands go in here
'''
```

2.8. Folder Structure for Analysis Projects

One big advantage of R markdowns is that when opening up an .Rmd file in RStudio, the working directory is always set to the location of the file. This facilitates reproducibility because somebody does not have to change the working directory to fit the folder structure of their own machine. It's generally frowned upon to use setwd() in a script for this reason, since any use of setwd() is specific to one's machine. Instead, wouldn't it be much easier if the user who wants to reproduce your analysis just needs to download the entire project folder, with no fiddling of setwd() required? R markdown scripts make this possible.

In general, it is good to structure your project in a consistent manner around folders. The R markdown scripts then operate *relatively* from within that folder structure, not *absolutely* specific to the folder structure on your machine. At a bare minimum, there should be a data folder, a scripts folder, and a figures folder (see Figure 2.4).

Let's say you are working within the 'analysis.Rmd' file in the 'scripts' folder, aiming to load the 'mydata.csv' file from the 'data' folder. The following command would achieve this:

```
mydf <- read_csv('../data/mydata.csv')
```

The '. .' instructs the markdown file to jump one folder up in the hierarchy of folders, which is the overarching 'project' folder in this case. From there, the '/data/'

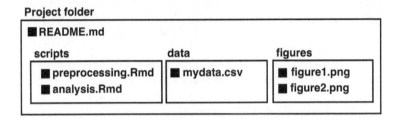

Figure 2.4. Folder structure for a data analysis project; black squares represent data files

bit instructs the markdown script to look for the 'data' folder. The logic is similar for saving a ggplot from within a script in the 'scripts' to the 'figures' folder:

```
ggsave('../figures/figure1.png')
```

2.9. Readme Files and More Markdown

The main project folder of any data analysis project should always contain a readme file describing the overall structure of the analysis. For this, markdown files (.md) are useful, as these files are often interpreted by data sharing repositories such as GitHub or the Open Science Framework (OSF). You can create markdown files with any text editor, such as the built-in 'Notes' on Macs or 'Notepad' on PCs.

Let's talk about some markdown features, which can also be used in R markdown files (.Rmd): Single hashtags '#' or double hashtags '##' display the corresponding text as a major or minor title on GitHub.

```
# My major title
## My minor title
```

Text enclosed by two stars '**' will be displayed in bold; text enclosed by one star '*' will be displayed in italics.

```
**bold text**
*italic text*
```

Lines beginning with hyphens will be displayed as bullet points.

```
- bullet point 1
- bullet point 2
```

Here's an example of what the beginning of a README.md file for a project could look like:

```
# Title of my statistical analysis

-  **Study design & data collection:** My friend
-  **Statistical Analysis:** Bodo Winter
```

```
-  **Date:** 24/12/18

## Libraries required for this analysis:

-  tidyverse
-  lme4

## Script files contained in this analysis:

-  preprocessing.Rmd : Getting the data into shape
-  analysis.Rmd : Linear mixed effects model analysis
```

Ideally, there's also a 'codebook' for every dataset which details the content of every column. In particular, it is useful if there is a full description of which values are contained within each column, such as, 'dialect: categorical variable, assumes one of the three values Brummie, Geordie, Scouse'.

It is important to use the features introduced in this section in R markdown files to give your analysis structure. For example, you can use '#' and '##' to highlight major and minor sections of your analysis.[6] As will be explained in the next section, code cleanliness is more than just cosmetics—it is intimately tied to the reproducibility of an analysis.

2.10. Open and Reproducible Research

Scientific progress is cumulative—it builds on past achievements. However, cumulative progress is only possible if results are both 'replicable' and 'reproducible'. What's the difference between replicability and reproducibility? In short: replicability characterizes the ability to replicate a study, that is, being able to conduct the same study again (with new data). Reproducibility characterizes the more basic requirement of being able to reproduce a researcher's analysis of a given dataset on one's own machine.

A study is replicable if another researcher can read the methods section of a paper and has enough information to replicate the study with new participants. Recently, researchers have found that many famous results failed to replicate, which has led to the 'replication crisis' (Open Science Collaboration, 2015; Nieuwland et al., 2018). Linguistics doesn't have a replication crisis yet, but it's looming around the corner. There already are important linguistic results that failed to replicate, such as the idea that there is a bilingual advantage in certain cognitive processing tasks (Paap & Greenberg, 2013; de Bruin, Bak, & Della Sella, 2015). Other linguistic findings that failed to replicate involve sentence processing (Nieuwland et al., 2018; Stack, James, & Watson, 2018) and embodied language understanding (Papesh, 2015).

There are many reasons why a study may not replicate. One reason, however, has to do with a lack of reproducibility of existing research. For any study, it should be possible for other researchers to obtain the same results if they were given the same data. The problem is that statistical analysis—as will be pointed out repeatedly throughout this book—is a strikingly subjective process (this may surprise you). For example,

---

6 Note that inside a code chunk, the hashtag '#' is interpreted as a comment. Outside of a code chunk, it is interpreted as a title.

even expert analysts will come to different conclusions when given the same data-set (Silberzahn et al., 2018). There are myriads of decisions to make in an analysis, what some people call "researcher degrees of freedom" (Simmons, Nelson, & Simon-sohn, 2011), and what others call "the garden of forking paths" (Gelman & Loken, 2014). Without the ability to trace what a researcher did in their analysis, these choices remain hidden. Reproducibility requires us to lay all of this open.

A fully reproducible study gives you the data and the code. When you execute the code on your machine, you will get exactly the same results. However, reproducibility is a gradable notion, with some studies being more reproducible than others. There are several things that increase a study's reproducibility:

- The minimal requirement for reproducible research is that the data and scripts are shared ('open data, open code'). The scientific community, including the linguistic community, needs to rethink what it means to 'publish' a result. The following mindset may be helpful to induce change and to assure replicability and reproducibility in the long run: without publishing the data and code, a publication is *incomplete*. The scientific process hasn't finished unless the data and code are provided as well.
- The choice of software influences reproducibility. R is more reproducible than SPSS, STATA, SAS, Matlab, or other proprietary software, because it's free, open-source, and platform-independent. Whenever possible, use software that is accessible to everybody.
- More thoroughly documented code is more reproducible, as other researchers will have an easier time tracing your steps.
- R markdown scripts facilitate reproducibility because they make extensive documentation easier via the ability to incorporate plain text, and because they allow avoiding `setwd()` (see discussion above). In addition, the ability to knit a final report of your analysis allows researchers (including yourself) to more quickly see the results *together* with the code that has produced the results.
- Use publicly accessible platforms such as OSF and GitHub[7] rather than a journal's online supplementary materials. It's not in the publisher's interest to store your data for eternity, and it's well known that publishers' websites are subject to change, which leads to link rot.

When people first hear about the idea of open reproducible research, they may be worried. There are a few common objections that are worth addressing at this stage (see also Houtkoop, Chambers, Macleod, Bishop, Nichols, & Wagenmakers, 2018).

- *I don't want to my results or ideas stolen (I don't want to get 'scooped').* This is a very common worry. Paradoxically, perhaps, making everything open from the get-go actually makes it *less* likely to get scooped. Publishing on a publicly

---

7 GitHub and OSF play well together. For my analysis projects, I use an OSF repository as the overarching repo, linking to a corresponding GitHub repo which stores the data and code. OSF will become more important in years to come, as more and more researchers are 'pre-registering' their studies (see Nosek & Lakens, 2014).

accessible repository allows you to claim precedence, which makes scooping more difficult.[8]

- *I fear retaliation or loss of reputation if somebody finds an error, or provides an alternative analysis with different conclusions.* This is a very common fear, and I think it's safe to say that most scientists have worried about this at some point in their career. However, this objection gets it the wrong way around. You are more likely to be criticized if somebody finds out that something is wrong and you have *not* shared your data and code. When making your materials openly accessible, people are less likely to ascribe deliberate wrongdoing to you. Keeping your data behind locked doors is only going to make you look more suspicious. You and your research will appear more trustworthy if you share everything right away.

- *I cannot actually share the data for ethical reasons.* Certain datasets are impossible to share for very good reasons. However, it's usually the case that the final steps of a data analysis can be shared, such as the summary data that was used to construct a statistical model. In general, it is your responsibility that the data is made anonymous so that it can be appropriately shared without ethical concerns.

- *I fear that companies may use my data for wrongdoing.* It is good to be concerned about big-data-harvesting that is done with bad intentions, especially when it involves the recognition of identities. That said, almost all of the data dealt with in linguistics is anonymous or, if not, it can easily be anonymized. Moreover, the data of most studies is often of little use to companies. In particular, most experimental studies are quite constrained to very particular, theoretically involved hypotheses that cannot easily be commercialized. Things are different if you are dealing with non-anonymized data, or such data sources as social media. However, in general, compared to the overwhelming benefits of sharing data and code in the face of the replication crisis, corporate misuse is a very minor concern. If you are truly worried about corporate data pirates, consider licensing your data accordingly.

- *I feel embarrassed about my code.* This is a very understandable concern—don't worry, as you're not alone! The first code that I put up online a few years ago looks absolutely horrible to me right now. People understand that everybody's coding practice is different, and that different researchers have different levels of statistical sophistication. Nobody will look at your code and laugh at it. Instead, they will appreciate the willingness to share your materials, even if it doesn't look snazzy.

- *Sharing data and code is not a common practice in my field.* My response to this is: not yet! Things are clearly heading this way, and more and more journals are implementing policies for data sharing—many already have! Journals such as *PLOS One* and the Royal Society journals are trendsetters in this respect. In addition: if it's not yet a common practice in your field, why not make it a common practice? You can lead the change.

Putting scientific progress aside, if you wanted some purely cynical reasons for 'going open' and reproducible, you might want to know that studies with open data and open code have higher citation rates (Piwowar & Vision, 2013).

---

8 Moreover, let's face it: many linguists are working on such highly specific topics that getting scooped is of little concern. However, if you are working on something that many other people are currently working on as well, use a publicly accessible repository to be the first to get it out there!

Finally, since all kinds of disciplines (including the language sciences) are currently moving towards open and reproducible methods, it makes sense for you to adopt open and reproducible methods early on, so that you are not overwhelmed when this becomes a general requirement. Rather than being late adopters, let's try to be ahead of the curve and at the same time further progress in science!

2.11. Exercises

2.11.1. Review

Review the tidyverse style guide ...

> http://style.tidyverse.org/

> ... and the RStudio keyboard shortcut list:

> https://support.rstudio.com/hc/en-us/articles/200711853-Keyboard-Shortcuts

Think about which shortcuts you want to adopt in your own practice (experiment!). After having incorporated a few shortcuts into your workflow, return to the shortcut list and think about which other shortcuts to adopt.

2.11.2. Create and Knit a Markdown File

Create a markdown file with the title 'Analysis of linguistic diversity'. In the first R code chunk, load in the tidyverse package and the Nettle (1999) dataset. Using the `sum()` function, compute the sum of languages across the entire dataset. Describe each step with a few short sentences of plain text outside of the code chunks. Then, knit the file to an html file and check the output.

2.11.3. Subsetting Data Frames with Tidyverse Function

This exercise uses the `nettle` data frame to explore different ways of indexing using `filter()` and `select()`. First, load in the `nettle` data:

```
nettle <- read.csv('nettle_1999_climate.csv')

head(nettle)   # display first 6 rows
```

```
  Country Population Area  MGS  Langs
1  Algeria       4.41 6.38 6.60     18
2   Angola       4.01 6.10 6.22     42
3 Australia      4.24 6.89 6.00    234
4 Bangladesh     5.07 5.16 7.40     37
5   Benin        3.69 5.05 7.14     52
6  Bolivia       3.88 6.04 6.92     38
```

Next, attempt to understand what the following commands do. *Then* execute them in R and see whether the output matches your expectations.

```
filter(nettle, Country == 'Benin')

filter(nettle, Country %in% c('Benin', 'Zaire'))

select(nettle, Langs)

filter(nettle, Country == 'Benin') %>% select(Langs)

filter(nettle, Country == 'Benin') %>%
select(Population:MGS)

filter(nettle, Langs > 200)

filter(nettle, Langs > 200, Population < median(Population))
```

2.11.4. Extended Exercise: Creating a Pipeline

Execute the following code in R (you may omit the comments for now) and then read the explanation below.

```
# Reduce the nettle tibble to small countries:

smallcountries <- filter(nettle, Population < 4)

# Create categorical MGS variable:

nettle_MGS <- mutate(smallcountries,
     MGS_cat = ifelse(MGS < 6, 'dry', 'fertile'))

# Group tibble for later summarizing:

nettle_MGS_grouped <- group_by(nettle_MGS, MGS_cat)

# Compute language counts for categorical MGS variable:

summarize(nettle_MGS_grouped, LangSum = sum(Langs))
```

```
# A tibble: 2 x 2
  MGS_cat LangSum
  <chr>     <int>
1 dry         447
2 fertile    1717
```

This code reduces the nettle tibble to small countries (Population < 4). The resulting tibble, smallcountries, is changed using the ifelse() function.

In this case, the function splits the dataset into countries with high and low ecological risk, using six months as a threshold. The ifelse() function spits out 'dry' when MGS < 6 is TRUE and 'fertile' when MGS < 6 is FALSE. Then, the resulting tibble is grouped by this categorical ecological risk measure. As a result of the grouping, the subsequently executed summarize() function knows that summary statistics should be computed based on this grouping variable.

This code is quite cumbersome! In particular, there are many intervening tibbles (smallcountries, nettle_MGS, and nettle_MGS_grouped) that might not be used anywhere else in the analysis. For example, the grouping is only necessary so that the summarize() function knows what groups to perform summary statistics for. These tibbles are fairly dispensable. Can you condense all of these steps into a single pipeline where the nettle tibble is first piped to filter(), then to mutate(), then to group_by(), and finally to summarize()?

3 Descriptive Statistics, Models, and Distributions

3.1. Models

This book teaches you how to construct statistical models. A model is a simplified representation of a system. For example, the map of a city represents a city in a simplified fashion. A map providing as much detail as the original city would not only be impossible to construct, it would also be pointless. Humans build models, such as maps and statistical models, to make their lives simpler.

Imagine having conducted a reading time experiment that involves measurements from 200 participants. If you wanted to report the outcome of your experiment to an audience, you wouldn't want to talk through each and every data point. Instead, you report a summary, such as 'The 200 participants read sentences with an average speed of 2.4 seconds', thus saving your audience valuable time and mental energy. This chapter focuses on such summaries of numerical information, specifically, the mean, the median, quantiles, the range, and the standard deviation. The mean is a summary of a distribution. What exactly is a distribution?

3.2. Distributions

Imagine throwing a single die 20 times in a row. For a regular die, there are six possible outcomes. For any one throw, each face is just as likely to occur as any other. Let's perform 20 throws and note down how frequently each face occurs. The face '1' comes up 2 times, '2' might come up 5 times, and so on. The result of tallying all counts is a 'frequency distribution', which associates each possible outcome with a particular frequency value. The corresponding histogram is shown in Figure 3.1a (for an explanation of histograms, see Chapter 1.12).

The distribution in Figure 3.1a is an empirically observed distribution because it is based on a set of 20 actual throws of a die ('observations'). Figure 3.1b on the other hand is a theoretical distribution, one that isn't attested in any specific dataset. Notice how the y-axis in Figure 3.1b represents probability rather than frequency. Probabilities range from 0 to 1, with 0 indicating that a given event never occurs and 1 indicating that a given event always occurs. The theoretical probability distribution in Figure 3.1b answers the question: how probable is each outcome? In this case, all outcomes are equally probable, namely, 1 divided by 6, which is about 0.17. In other

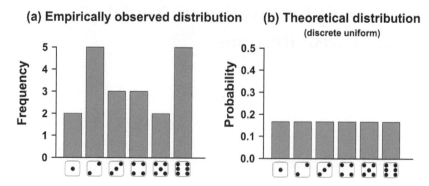

Figure 3.1. (a) An empirically observed distribution based on 20 throws of a die; (b) A theoretical distribution displaying the expected probabilities for an infinite number of throws

words, each face is expected to occur one sixth of the time, although any particular set of throws may deviate from this theoretical expectation.

Commonly used theoretical distributions have names. The particular distribution shown in Figure 3.1b is the 'discrete uniform distribution'. It is a 'uniform' distribution because the probability is uniformly spread across all possible outcomes. It is furthermore a 'discrete' distribution because there are only six particular outcomes and no in-betweens.

Applied statistics involves both empirical distributions and theoretical distributions. The theoretical distributions are tools that help modeling empirically observed data. Most models constructed in applied statistics *assume* that the data has been generated by a process following a certain distribution. In order to model various types of data, you have to learn about various theoretical distributions. This chapter introduces the normal distribution. Later chapters introduce the Bernoulli distribution (Chapter 12) and the Poisson distribution (Chapter 13).

3.3. The Normal Distribution

One of the most common distributions in statistics is the 'normal distribution', also known as the 'bell curve' due to its characteristic bell shape. A more technical name for this distribution is the Gaussian distribution, after the mathematician Carl Friedrich Gauss. The normal distribution is a distribution for continuous data, centered symmetrically around the mean with the bulk of data lying close to the mean.

Figure 3.2a shows three distributions of actual data that are approximately normally distributed. To make this example more concrete, you can imagine that these are language test scores from three different classrooms. Each of the three distributions has a different mean. Classroom A has a mean of 10 (this class performed badly overall), B has a mean of 40, and C has a mean of 80 (this class performed very well). In such a scenario, you can think of the mean as specifying the 'location' of

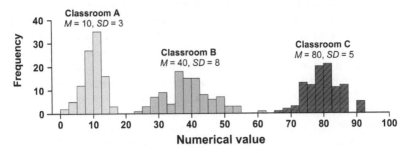

(a) Distributions with differing means and standard deviations

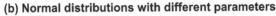

(b) Normal distributions with different parameters

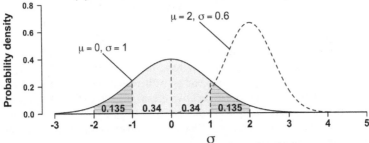

Figure 3.2. (a) Three distributions for groups of students with 100 students each; the data is random data that was generated based on an underlying normal distribution with the specified means and standard deviations. (b) Two normal distributions with different parameters; the parameter μ ('mu') specifies the location of the distribution on the number line; σ ('sigma') specifies the spread; the highlighted areas under the curve show the 68% and 95% intervals

a distribution on the *x*-axis. That is, the mean tells you how large or small a set of numbers is overall.[1]

The distributions in Figure 3.2a also differ in terms of 'spread'. In particular, the distribution for classroom B is wider than the other two distributions. This means that there are more students farther away from the mean in this classroom. Students in this classroom are more different from each other than the students in the other

---

1 The notation $\bar{x}$ ('x bar') is often used to represent the mean of a set of numbers *x*. The formula for the mean is $\bar{x} = \dfrac{\sum x}{N}$. Formulas are summary formats for computational procedures. The $\sum x$ in the numerator stands for the procedure that sums up all the numbers. The sum is then divided by N, which represents how many numbers there are. For example, imagine collecting three response durations in a psycholinguistic experiment, specifically 300*ms*, 200*ms*, and 400*ms*. The sum of these numbers is 900*ms*, with $N = 3$, which yields $\dfrac{900ms}{3} = 300ms$.

classrooms. The standard deviation (*SD*) is used to summarize this spread. Although the actual formula has some quirks,[2] it is OK for the most part to think of the standard deviation as representing the average distance from the mean. Visually, larger standard deviations correspond to flatter histograms that fan out farther away from the mean.

While the mean and standard deviation can be computed for any data, these two numbers have special meaning in the context of the normal distribution: they are the distribution's 'parameters'. A parameter is a property of a distribution. Changing the parameter changes the look of the distribution. Changing the mean moves the normal distribution along the number line, and changing the standard deviation stretches or squeezes the distribution.

Figure 3.2b shows the normal distribution for two different sets of parameters. The *y*-axis label says 'probability density', which means that the height of the graph indicates how probable certain values are. The area under each bell curve adds up to 1.0. You may also have noticed that I tacitly switched notation. Before, I used Roman letters to represent the mean (*M* or $\bar{x}$) and the standard deviation (*SD* or *s*). In Figure 3.2b, I used the Greek letters μ ('mu') and σ ('sigma') instead. It is conventional to use Greek letters when talking about the parameters of theoretical distributions. When talking of empirically observed data, it is conventional to use Roman letters instead. This is not just an arbitrary convention—you will later see that this notation is quite important.

Figure 3.2b also shows an important property of the normal distribution: the areas highlighted with '0.34' add up to a probability of 0.34, and so on. Together, the two middle parts add up to $p = 0.68$. If you were to randomly draw numbers from this distribution, 68% of the time, you would end up with a number that is between -1 and $+1$ standard deviations. If you add the next two areas (striped) to this, the probability mass under the curve adds up to $p = 0.95$. Thus, if you were to draw random data points from this distribution, 95% of the time you would end up with a number that is between -2 and $+2$ standard deviations.

The 68%–95% 'rule' allows you to draw a mental picture of a distribution from the mean and standard deviation alone, granted the distribution is approximately normally distributed. Let us apply the rule to a dataset from Warriner, Kuperman, and Brysbaert (2013), who asked native English speakers to rate the positivity or negativity of words on a scale from 1 to 9 ('emotional valence'). In their paper, the authors report that the mean of the ratings was 5.06, with a standard deviation of 1.27. Assuming normality, you can expect 68% of the data to lie between 3.79 and 6.33 ($5.06 - 1.27$, $5.06 + 1.27$). You can furthermore expect 95% of the data to lie between 2.52 and 7.6. To calculate

---

2 The formula for the standard deviation requires that you calculate the mean first, as the standard deviation measures the spread from the mean. The formula works as follows: you calculate each data point's difference from the mean, square that value, and subsequently sum up the squared differences. This 'sum of squares' is then divided by *N* minus *1* (so *99*, if you have *100* data points). Subsequently, you take the square root of this number, which yields the standard deviation. There are reasons for why the square root has to be used and why one has to divide by $N - 1$ (rather than by *N*) that I will not go into here. The abbreviated formula for the standard deviation of a set of numbers, *x*, is: $SD = \sqrt{\dfrac{\sum(x-\bar{x})^2}{n-1}}$. When you see a formula like this, it is important to try to understand it piece by piece, and to first focus on those symbols that you recognize. For example, you may observe in this formula that there's an $\bar{x}$, which represents the mean, and that this number is subtracted from each *x*-value. Mathematicians use formulas not to make things more difficult, but to make things *easier* for them. A formula encapsulates a lot of information in a neat way.

this, you have to double the standard deviation 1.27, which yields 2.54. Then you add and subtract this number from the mean (5.06 − 2.54, 5.06 + 2.54).

Finally, it is worth noting that, because the mean and standard deviation work so nicely together, it's generally a good idea to report both. In particular, means without a measure of spread are not very informative.

3.4. Thinking of the Mean as a Model

I invite you to think of the mean as a model of a dataset. For one, this highlights the compressive nature of the mean (given that models are simplified representations). Second, it highlights that the mean is a representation of something, namely, a distribution.

Moreover, thinking of the mean as a model highlights that the mean can be used to make predictions. For example, Warriner et al. (2013) rated 'only' 14,000 English words, even though there are many more words in the English language. The word *moribund* is one of the words that has not been rated. In the absence of any information, can we predict its emotional valence value? Our best guess for this word's value is the mean of the current sample, 5.06. In this sense, the mean allows making predictions for novel words.

Introductory statistics courses often distinguish between 'descriptive statistics' and 'inferential statistics'. Whereas descriptive statistics is understood to involve things like computing summary statistics and making plots, inferential statistics is generally seen as those statistics that allow us to make 'inferences' about populations of interest, such as the population of all English speakers, or the 'population' of all English words. However, the distinction between descriptive and inferential statistics is not as clear-cut. In particular, any description of a dataset can be used to make inferences. Moreover, all inferential statistics are based on descriptive statistics.

In fact, in some sense, you have already performed some form of inferential statistics in this chapter. The Warriner et al. (2013) dataset can be treated as a *sample* of words that is taken from the *population* of all English words. In applied statistics, we almost always deal with samples as the population is generally not available to us. For example, it may be infeasible to test all English speakers in a psycholinguistic experiment and hence we have to resort to a small subset of English speakers to estimate characteristics of the population.

Samples are used to *estimate* population parameters.[3] The distinction between sample estimates and population parameters is enshrined in mathematical notation. As mentioned above, parameters are conventionally represented with Greek letters; sample estimates, with Roman letters. Then, one can say that the sample mean $\bar{x}$ estimates the population parameter μ. Similarly, the sample standard deviation s estimates the population parameter σ. Other texts may use the caret symbol for this distinction, in which case $\hat{\mu}$ estimates μ, and $\hat{\sigma}$ estimates σ.

From now on, whenever you see means and standard deviations in published papers, ask yourself questions such as the following. What is this mean estimating? What is the relevant population of interest? In later chapters, you will quantify the degree of uncertainty with which sample estimates reflect population parameters.

---

3 This book is focused on a branch of statistics that is called 'parametric statistics'. Just so you've heard about it: there is a whole other branch of statistics that deals with 'non-parametric statistics', which as the name suggests, does not estimate parameters.

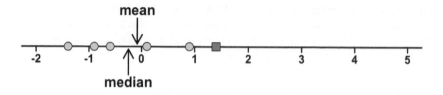

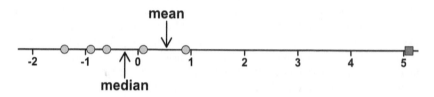

Figure 3.3. Six data points on the number line. The data point represented by the square shifts between the two datasets, pulling the mean upwards. The median stays put because the fact that 50% of the data is on either side hasn't changed by altering the value.

3.5. Other Summary Statistics: Median and Range

The mean and the standard deviation are just two ways of compressing data. Another summary statistic is the median, which is the halfway point of the data (50% of the data are above the median; 50% of the data are below). In contrast, the mean is the 'balance point', which you can think of as a pair of scales: an extreme value shifts the mean, just as a heavy object tips a pair of scales. This is exemplified in Figure 3.3.[4]

The fact that the median *doesn't* move when the position of the square is changed in Figure 3.3 can be seen as an advantage, as well as a disadvantage. The mean incorporates more information because it cares about the actual values of the data points. The median only cares about its position in an ordered sequence.

The median is sometimes reported because it is more robust to variations in extreme values than the mean. For example, most people have relatively low incomes, but some people (such as Bill Gates) have incredibly high incomes. Such extremely rich people skew the mean upwards, but they don't shift the median up as much.

Another summary statistic for the spread is the range, which is the distance between the smallest and the largest data point. For the Warriner et al. (2013) dataset, subtracting the minimum (1.26) from the maximum (8.53) yields the range 7.27. As a general measure of spread, the range is not as useful because it exclusively relies on the two most extreme numbers, ignoring all others. If one of these numbers happens

---

4 For an uneven number of data points, there is a true middle number. However, for an even number of data points, there are two numbers in the middle. What is the halfway point in this case? The median is defined to be the mean of the two middle numbers, e.g., for the numbers 1, 2, 3, and 4, the median is 2.5.

to change, the range will change as well. However, many times it is useful to know specifically what the smallest and the largest number in a dataset are.

3.6. Boxplots and the Interquartile Range

Now that you have a better understanding of the median, let's return to the boxplot, which was briefly introduced but not explained in Chapter 2. Box-and-whisker plots are very common in many areas of science, and you will find them in many linguistic papers as well. To understand the meaning of a boxplot, you need to learn about the 'interquartile range'.

Figure 3.4 shows the distribution of emotional valence scores for the 14,000 words from the Warriner et al. (2013) rating study. Recall that the sample mean of this distribution is $M = 5.06$. The median is 5.2 and is shown as the thick black line in the middle of the box of the boxplot. The extent of the box covers 50% of the data, that is, 25% of the data above and below the median. The ends of the box have specific names: they are the first, second, and third 'quartile'. You are probably more familiar with percentiles. For example, if someone received a test score that is in the 80th percentile, 80% of the test scores are below that person's score. Q1 is the first quartile, which is the 25th percentile. The next quartile is Q2, the median. Finally, 75% of the data fall below Q3, the third quartile.

For the Warriner et al. (2013) dataset, Q1 is the number 4.25 (25% of all data points fall below this value), and Q3 is 5.95 (75% fall below this value). The difference between Q1 and Q3 yields the 'interquartile range', in this case, $5.95 - 4.25 = 1.7$. This number corresponds to the length of the box seen in Figure 3.4.

What's the meaning of the whiskers extending from the box? Let's focus on the right whisker, the one that is extending from Q3 towards larger emotional valence scores. This whisker ends at the largest number that falls within a distance of 1.5 times the

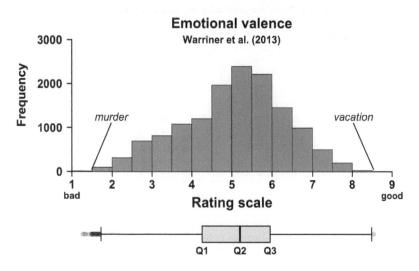

Figure 3.4. A histogram of the emotional valence rating data

interquartile range from Q3. You can think of standing at Q3 and swinging a lasso as wide as 1.5 times the interquartile range. The largest number that you catch—the score for the word *happiness* in this case—is the extent of the whisker. The logic is the same for the lower whisker, except that you are looking for the smallest number that falls within a distance of $1.5 * IQR$.

The data points that fall outside of the whiskers are indicated by dots. For example, the single dot to the right of the right whisker is the word *vacation*. Words that fall outside the range of the whiskers are often called outliers, but I prefer the term 'extreme value', since 'outlier' suggests that something is qualitatively different from the other data points, which is often used to justify exclusions.[5] Using the term 'extreme value' implies that the same underlying process has generated the extremity.

If 'maximum of $Q3 + 1.5 * IQR$' and 'minimum of $Q1 - 1.5 * IQR$' seems like a horribly non-intuitive way of defining the whiskers to you, perhaps it's best to avoid boxplots. And, if you do decide to use a boxplot, don't hesitate to re-state the definition of the whiskers in the figure captions—it's good to give people reminders.[6] Describing the meaning of the individual plot components in the figure captions should be done anyway. For example, sometimes researchers use the range of the data (minimum and maximum) for the whiskers of a boxplot, which you would need to know in order to interpret the plot correctly. In general, when writing up statistical results, it's good if you describe each plot in as much detail as possible.

3.7. Summary Statistics in R

Let's put our understanding of distributions and summary statistics into practice. First, create 50 uniformly distributed numbers with the runif() function. The name of this function stands for 'random uniform'. Since this is a random number generation function, your numbers will be different from the ones shown in this book.

```
# Generate 50 random uniformly distributed numbers:

x <- runif(50)

# Check:

x
```

```
 [1]  0.77436849 0.19722419 0.97801384 0.20132735
 [5]  0.36124443 0.74261194 0.97872844 0.49811371
 [9]  0.01331584 0.25994613 0.77589308 0.01637905
[13]  0.09574478 0.14216354 0.21112624 0.81125644
[17]  0.03654720 0.89163741 0.48323641 0.46666453
```

5 Data should never be excluded unless there are justifiable reasons for doing so.
6 I suspect that many people in the language sciences may not be able to state the definition of the whiskers off the top of their heads. If you want to make new friends at an academic poster session, next time you spot a boxplot, ask the poster presenter to define the whiskers.

```
[21] 0.98422408 0.60134555 0.03834435 0.14149569
[25] 0.80638553 0.26668568 0.04270205 0.61217452
[29] 0.55334840 0.85350077 0.46977854 0.39761656
[33] 0.80463673 0.50889739 0.63491535 0.49425172
[37] 0.28013090 0.90871035 0.78411616 0.55899702
[41] 0.24443749 0.53097066 0.11839594 0.98338343
[45] 0.89775284 0.73857376 0.37731070 0.60616883
[49] 0.51219426 0.98924666
```

Notice that by default, the `runif()` function generates continuous random numbers within the interval 0 to 1. You can override this by specifying the optional arguments `min` and `max`.

```
x <- runif(50, min = 2, max = 6)

head(x)
```

```
[1] 2.276534 2.338483 2.519782 4.984528 2.155517
[6] 4.742542
```

Plot a histogram of these numbers using the `hist()` function. A possible result is shown in Figure 3.5 (left plot). Remember that your plot will be different, and that is OK.

```
hist(x, col = 'steelblue')
```

Next, generate some random normally distributed data using the `rnorm()` function and draw a histogram of it.

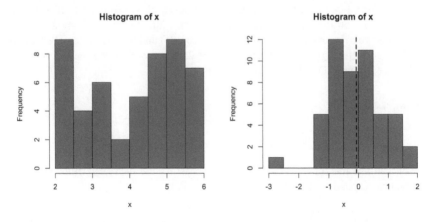

Figure 3.5. Left: random data drawn from aa uniform distribution; right: random data drawn from a normal distribution; the dashed line indicates the mean

```
x <- rnorm(50)

hist(x, col = 'steelblue')

abline(v = mean(x), lty = 2, lwd = 2)
```

This code also plots a vertical line at the mean. The line type is indicated to be dashed (`lty = 2`), and the line width is indicated to be 2 (`lwd = 2`). Figure 3.5 (right plot) is an example distribution.

The `rnorm()` function generates data with a mean of 0 by default. It also has a default for the standard deviation, which is 1 . You can override these defaults by specifying the optional arguments `mean` and `sd`.

```
x <- rnorm(50, mean = 5, sd = 2)
```

Check whether this has worked by computing the mean and the standard deviation using the corresponding `mean()` and `sd()` functions (remember: your values will be different).

```
mean(x)
```

```
[1] 4.853214
```

```
sd(x)
```

```
[1] 2.262328
```

Notice that these values are close to what was specified in the `rnorm()` command (mean = 5, sd = 2).

The `quantile()` function is useful for computing percentiles. If you run `quantile()` on the vector of random numbers without supplying any additional arguments, you retrieve the minimum (0th percentile) and the maximum (100th percentile), as well as Q1 (the first 'quartile', the 25th percentile), the median (Q2, the 50th percentile) and Q3 (the 75th percentile).

```
quantile(x)
```

```
       0%       25%       50%       75%      100%
-1.297574  3.100322  4.633111  6.363569 10.157849
```

You can use the `quantile()` function to assess the 68%-95% rule. The 68% interval corresponds to the 16th and 84th percentiles.

```
quantile(x, 0.16)
```

```
     16%
2.749623
```

```
quantile(x, 0.84)
```

```
     84%
7.080644
```

If the 68% rule of thumb works, the resulting numbers should be fairly close to the interval covered by $M - SD$ and $M + SD$.

```
mean(x) - sd(x)
```

```
[1] 2.590886
```

```
mean(x) + sd(x)
```

```
[1] 7.115541
```

And, indeed, the numbers are fairly similar to the percentiles. Let's do the same for the 95% interval, which corresponds to the interval between the 2.5th and the 97.5th percentiles. For this small example dataset, the 95% rule is a little off.

```
# 2.5th percentile:
```

```
quantile(x, 0.025)
```

```
    2.5%
1.004807
```

```
# Should correspond to M - 2 * SD:
```

```
mean(x) - 2 * sd(x)
```

```
[1] 0.3285584
```

```
# 97.5th percentile:
```

```
quantile(x, 0.975)
```

```
   97.5%
8.758132
```

```
# Should correspond to M + 2 * SD:
```

```
mean(x) + 2 * sd(x)
```

```
[1] 9.377869
```

I highly recommend using random number generation functions to develop an intuition for how approximate Gaussian data looks like. In some circumstances, a histogram may look quite non-Gaussian even though an underlying Gaussian distribution was used to generate the data. To get a 'feel' for drawing random samples from the normal distribution, execute the following command repeatedly.

```
hist(rnorm(n = 20))   # execute repeatedly
```

You will notice that with only 20 data points it is often quite difficult to see that the data was drawn from an underlying normal distribution. If you change n to a very large number, the histograms should be much more Gaussian in shape. The take-home message here is that it is often difficult to tell what the appropriate reference distribution is for very small sample sizes.

3.8. Exploring the Emotional Valence Ratings

In this section, you will analyze the above-mentioned emotional valence ratings from Warriner et al. (2013). Let's load the tidyverse package and the data into your current R session (you will have to make sure that your working directory is set appropriately; see Chapter 1).

```
# Load packages and data:

library(tidyverse)

war <- read_csv(' warriner_2013_emotional_valence.csv')
```

As was emphasized again and again in Chapters 1 and 2, whenever you load data into R, you should spend considerable time familiarizing yourself with its structure.[7]

```
war
```
```
# A tibble: 13,915 x 2
    Word         Val
    <chr>        <dbl>
 1 aardvark     6.26
 2 abalone      5.3
 3 abandon      2.84
 4 abandonment  2.63
 5 abbey        5.85
 6 abdomen      5.43
 7 abdominal    4.48
 8 abduct       2.42
 9 abduction    2.05
10 abide        5.52
# ... with 13,905 more rows
```

---

7 It's worth remembering the principle 'garbage in, garbage out': if there's some issue with the data that you are unaware of, any stats computed may be worthless.

The tibble has two columns, pairing a `Word` with an emotional valence rating, `Val`. Compute the range to get a feel for this measure:

```
range(war$Val)
```

```
[1] 1.26 8.53
```

The emotional valence scores range from 1.26 to 8.53. To find the corresponding words, use `filter()`. Remember that this function *filters* rows based on a logical condition (see Chapter 2).

```
filter(war, Val == min(Val) | Val == max(Val))
```

```
# A tibble: 2 x 2
  Word        Val
  <chr>       <dbl>
1 pedophile   1.26
2 vacation    8.53
```

The above command uses the logical function 'or' (represented by the vertical bar '|') to retrieve all the rows that satisfy either of the two logical statements. This command can be translated into the following English sentence: 'Filter those rows from the `war` tibble for which valence is equal to the minimum or the maximum (both are OK).' The following command achieves the same result in a more compressed fashion (see Chapter 2.6 for an explanation of `%in%`).

```
filter(war, Val %in% range(Val))
```

```
# A tibble: 2 x 2
  Word        Val
  <chr>       <dbl>
1 pedophile   1.26
2 vacation    8.53
```

Let's have a look at the most positive and the most negative words in the dataset by using `arrange()`.

```
arrange(war, Val)   # ascending order
```

```
# A tibble: 13,915 x 2
  Word        Val
  <chr>       <dbl>
1 pedophile   1.26
2 rapist      1.30
3 AIDS        1.33
4 torture     1.40
5 leukemia    1.47
6 molester    1.48
7 murder      1.48
```

```
 8 racism       1.48
 9 chemo        1.50
10 homicide     1.50
# ... with 13,905 more rows
```

```
arrange(war, desc(Val))    # descending order
```

```
# A tibble: 13,915 x 2
   Word        Val
   <chr>       <dbl>
 1 vacation    8.53
 2 happiness   8.48
 3 happy       8.47
 4 christmas   8.37
 5 enjoyment   8.37
 6 fun         8.37
 7 fantastic   8.36
 8 lovable     8.26
 9 free        8.25
10 hug         8.23
# ... with 13,905 more rows
```

Thanks to surveying lots of different data points, you now have a firm grasp of the nature of this data. Let's compute the mean and standard deviation.

```
mean(war$Val)
```

```
[1] 5.063847
```

```
sd(war$Val)
```

```
[1] 1.274892
```

The mean is 5.06; the standard deviation is 1.27. You expect 68% of the data to follow into the following interval:

```
mean(war$Val) - sd(war$Val)
```

```
[1] 3.788955
```

```
mean(war$Val) + sd(war$Val)
```

```
[1] 6.338738
```

Verify whether this is actually the case using the quantile() function. The fact that the resulting numbers are close to $M - SD$ and $M + SD$ shows that the rule was pretty accurate in this case.

```
quantile(war$Val, c(0.16, 0.84))
```

```
16%  84%
3.67 6.32
```

Finally, let's have a look at the median, which is very similar to the mean in this case.

```
median(war$Val)
```

```
[1] 5.2
```

```
quantile(war$Val, 0.5)
```

```
50%
5.2
```

3.9. Chapter Conclusions

Everything in statistics is grounded in the notion of a distribution, and in (parametric) statistical modeling our goal is to make models of distributions. The mean is a great summary of a distribution, especially if the distribution approximates normality. In the applied R exercise, you then generated some random data and computed summary statistics. Being able to generate random data is a very important skill that will be nurtured throughout this book, alongside working with real data. Finally, you computed summary statistics for the Warriner et al. (2013) emotional valence ratings.

Everything up to this point has dealt with 'univariate' distributions. That is, you always only considered one set of numbers at a time. The next chapter will progress to bivariate data structures, focusing on the relationship between two sets of data.

3.10. Exercises

3.10.1. Exercise 1: Plotting a Histogram of the Emotional Valence Ratings

With the Warriner et al. (2013) data, create a ggplot2 histogram and plot the mean as a vertical line into the plot using geom_vline() and the xintercept aesthetic (see Chapter 2). Can you additionally add vertical dashed lines to indicate where 68% and 95% of the data lie? (Ignore any warning messages about binwidth that may arise).

3.10.2. Exercise 2: Plotting Density Graphs

In the plot you created in the last exercise, exchange geom_histogram() with geom_density(), which produces a kernel density graph. This is a plot that won't be covered in this book, but by looking at it you may be able to figure out that it is essentially a smoothed version of a histogram. There are many other geoms to explore. Check out the vast ecosystem of online tutorials for different types of ggplot2 functions.

Additional exercise: set the `fill` argument of `geom_density()` to a different color (such as 'peachpuff'). This is not an aesthetic mapping, because it doesn't draw from the data.

3.10.3. Exercise 3: Using the 68%-95% to Interpret Research Papers

Imagine reading a research paper about a grammaticality rating study. It is noted that the mean acceptability rating for a particular grammatical construction is 5.25 with a standard deviation of 0.4. Assuming normality, what is the interval within which you expect 68% of the data to lie? What about 95% of the data? Do you think the assumption of approximate normality is reasonable in this case?

4   Introduction to the Linear Model

Simple Linear Regression

4.1. Word Frequency Effects

The last chapter focused on modeling distributions with means. This chapter teaches you how to condition a mean on another variable. That is, you will create models that predict *conditional means*—means that shift around depending on what value some other piece of data assumes. In modeling such conditional means, you move from the topic of univariate statistics (describing single variables) to bivariate statistics (describing the relationship between two variables). The approach you learn in this chapter is the foundation for everything else in the book.

Hundreds of studies have found that frequent words are comprehended faster than infrequent words (e.g., Postman & Conger, 1954; Solomon & Postman, 1952; Jescheniak & Levelt, 1994). Figure 4.1 is an example of this, displaying response durations from a psycholinguistic study conducted as part of the English Lexicon Project (Balota et al., 2007). The *y*-axis extends from 400*ms* (two fifths of a second) to 1000*ms* (one second).[1] Longer response durations (up on the graph) mean that participants responded more slowly; shorter response durations (down on the graph) mean they responded faster. The word frequencies on the *x*-axis are taken from the SUBTLEX corpus (Brysbaert & New, 2009). They are represented on a logarithmic scale ($\log_{10}$). Logarithms will be explained in more detail in Chapter 5. From now on I will simply talk of 'word frequency' instead of 'log frequency', as for understanding the basics of regression models the logarithmic nature of the scale is not relevant.

The relationship between response duration and frequency is neatly summarized by a line. This line is the regression line, which represents the average response duration for different frequency values. Simple linear regression is an approach that models a single continuous response variable as a function of a predictor variable. Table 4.1 shows some common terminological differences that come up in regression modeling. Some researchers and textbooks use the language of 'regressing *y* on *x*'. I prefer to speak of 'modeling *y* as a function of *x*'. Researchers often use the terms 'independent variable' and 'dependent variable' (the dependent variable *y depends* on the

---

1   In what is called a 'lexical decision task', participants were asked whether a word was English or not. For example, *horgous* is not an English word, but *kitten* is. In Figure 4.1, each point is averaged over multiple participants' responses.

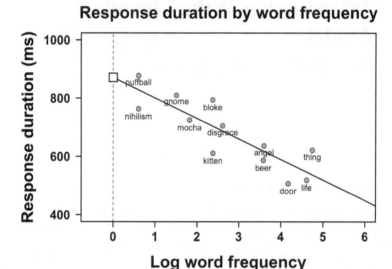

Figure 4.1. Response duration as a function of word frequency; each point represents the average response duration for a particular word; the *x*-axis is scaled logarithmically (see Chapter 5); the line represents the linear regression fit; the white square represents the intercept (the point where $x = 0$)

Table 4.1. Different names for response and predictors

y	x
response/outcome dependent variable	predictor independent variable explanatory variable regressor

independent variable *x*). I prefer to speak of *x* as a 'predictor', as it allows forming predictions for *y*, which I often call the 'response' or 'outcome' variable.

Generally, you specify regression in the direction of assumed causality (e.g., word frequencies affect response durations rather than the other way around). However, it is important to remember the slogan 'correlation is not causation', as the regression model cannot tell you whether there actually is a causal relationship between *x* and *y*.[2]

---

2 For some delightful examples of this principle in linguistics, see Roberts and Winters (2013). For example, cultures taking siestas speak languages with reduced morphological complexity.

4.2. Intercepts and Slopes

Mathematically, lines are represented in terms of intercepts and slopes. Let's talk about slopes first, then intercepts. In the case of the frequency effect shown in Figure 4.1, the slope of the line is negative. As the frequency values increase, response durations decrease. On the other hand, a positive slope goes 'up' (as x increases, y increases as well). Figure 4.2a shows two slopes that differ in sign. One line has a slope of +0.5, the other one has a slope of –0.5.

The slope is defined as the change in y (Δy, 'delta y') over the change in x (Δx, 'delta x').

$$slope = \frac{\Delta y}{\Delta x} \qquad\qquad (E4.1)$$

Sometimes, the slogan 'rise over run' is used as a mnemonic for this calculation. How much do you have to 'rise' in y for a specified 'run' along the x-axis? In the case of the response duration data discussed in this chapter, the slope turns out to be $-70\frac{ms}{freq}$. Thus, for each increase in frequency by 1 unit, the predicted response duration decreases by 70ms. For a two-unit increase in word frequency, the predicted response duration decreases by 140ms ($= 70ms * 2$), and so on.

However, a slope is not enough to specify a line. For any given slope, there is an infinity of possible lines. Figure 4.2b shows two lines with the same slope but differing 'intercepts'. You can think of the intercept informally as the point 'where the line starts' on the y-axis. As the y-axis is located at $x = 0$, this means that the intercept is the predicted y-value for $x = 0$. For the data shown in Figure 4.1, this happens to be the

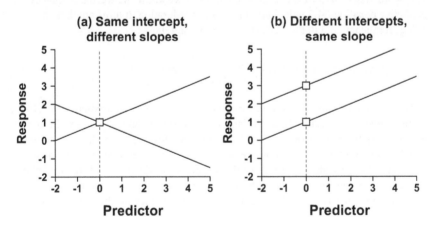

Figure 4.2. (a) Two lines with positive and negative slopes that go through the same intercept; (b) two lines with the same positive slope that have different intercepts

number 880ms (represented by the white square). Thus, for a (log) word frequency of 0, the model predicts the intercept 880ms.

Once the intercept and slope are fixed, there can be only one line. In math-speak, the line is said to be 'uniquely specified' by these two numbers. The intercept and slope are both 'coefficients' of the regression model. The letters b_0 and b_1 are commonly used to represent the intercept and the slope. Thus, the regression line has the following form:

$$y = b_0 + b_1 * x \tag{E4.2}$$

This equation yields different *y* means for different *x*-values. Let's plug in the actual coefficients from the regression shown in Figure 4.1.

$$response\ duration = 880ms + \left(-70\frac{ms}{freq}\right) * word\ frequency \tag{E4.3}$$

Because the slope has the unit $\frac{ms}{freq}$, multiplying it by a frequency value returns milliseconds. That is, $\frac{ms}{freq} * freq = ms$ (the two frequency units cancel each other out), highlighting how this regression model predicts response durations.

The coefficient estimates are the principal outcome of any regression analysis. A lot of your time should be spent on interpreting the coefficients—for example, by plugging in various values into the equation of your model to see what it predicts.

4.3. Fitted Values and Residuals

Let's see what response duration the regression model predicts for the word *script*. This word wasn't part of the original data, but if we know the word's frequency, the predictive equation E4.3 will churn out a response duration. It turns out that *script* has a (log) frequency of 3 in the SUBTLEX corpus, and thus:

$$response\ duration = 880ms + \left(-70\frac{ms}{freq}\right) * 3\ freq = 670ms \tag{E4.4}$$

The expected response duration for *script* is 670ms. Such a prediction is called a 'fitted value', as it results from 'fitting' a regression model to a dataset. In fact, all points along the regression line are fitted values. However, not all of these values may be equally sensible. Forming predictions based on a regression model generally only makes sense within the range of the attested data (this is called 'interpolation'). Regression models may produce odd results when predicting beyond the range of the attested data, what is called 'extrapolating'.[3] Word frequencies below zero make no

---

3 See Tatem, Guerra, Atkinson, and Hay (2004) for a hilarious extrapolation error. Based on the fact that female sprinters have increased their running speed more quickly than male sprinters over

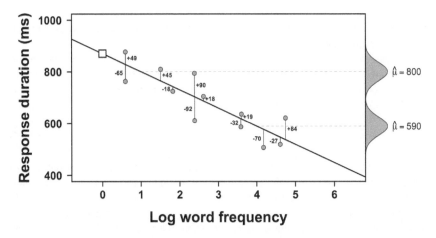

Figure 4.3. Regression line with vertical line segments indicating the residuals, which are positive (above the line) or negative (below the line); conditional means for $x = 1$ and $x = 4$ are projected onto the right y-axis; the density curves represent the constant variance and normality assumptions

sense, nevertheless, the regression model happily allows us to form predictions for negative values. For example, for the impossible x-value of -100, the model predicts a response duration of 7880ms. The model doesn't 'know' that negative frequencies make no sense.

The frequency model doesn't fit any of the data points perfectly. The extent to which the model is wrong for any specific data point is quantified by the residuals, which are the vertical distances of the observed data from the regression line, as shown in Figure 4.3. In this scatterplot, observed values above the line have positive residuals (+); observed values below the line have negative residuals (–). The actual numerical values represent how much the predictions would have to be adjusted upwards or downwards to reach each observed value.

The relationship between fitted values, observed values, and residuals can be summarized as follows:

$$observed\ values = fitted\ values + residuals \tag{E4.5}$$

This equation can be rewritten the following way:

$$residuals = observed\ values - fitted\ values \tag{E4.6}$$

---

the past few decades, these researchers claimed that, in 2156, women will run faster than men. As pointed out by Rice (2004) and Reinboud (2004), the same regression model would also predict instantaneous sprints in the year 2636.

Equation E4.6 shows that the residuals are what's 'left over' after subtracting your model's predictions from the data. Zuur, Ieno, Walker, Saveliev, and Smith (2009: 20) say that "the residuals represent the information that is left over after removing the effect of the explanatory variables".

Now that you know about residuals, the general form of a regression equation can be completed. All that needs to be done is to extend the predictive equation E4.2 with an 'error term', which I represent with the letter '*e*' in the equation below. In your sample, this term corresponds to the residuals.

$$y = b_0 + b_1 * x + e \tag{E4.7}$$

Essentially, you can think of the regression equation as being composed of two parts. One part is 'deterministic' and allows you to make predictions for conditional means (a mean dependent on *x*). This is the ' $b_0 + b_1 * x$ ' part of the above equation. This part is deterministic in the sense that for any value of *x* that you plug in, the equation will always yield the same result. Then, there is a 'stochastic' part of the model that messes those predictions up, represented by *e*.

4.4. Assumptions: Normality and Constant Variance

Statistical models, including regression, generally rely on assumptions. All claims made on the basis of a model are contingent on satisfying its assumptions to a reasonable degree. Assumptions will be discussed in more detail later. Here, the topic will be introduced as it helps us to connect the topics of this chapter, regression, with the discussion of distributions from the last chapter.

For regression, the assumptions discussed here are actually about the error term *e*, that is, they relate to the *residuals* of the model. If the model satisfies the normality assumption, its residuals are approximately normally distributed. If the model satisfies the constant variance assumption (also known as 'homoscedasticity'), the spread of the residuals should be about equal while moving along the regression line.

A clear violation of the constant variance assumption is shown in Figure 4.4a. In this case, the residuals are larger for larger *x*-values. Figure 4.4b demonstrates a clear violation of the normality assumption. In this case, a histogram of the residuals reveals 'positive skew', i.e., there are a few infrequent extreme values (see Chapter 5 for a discussion of skew). It is important to emphasize that the normality assumption is not about the response but about the residuals. It is possible that a model of a skewed response measure has normally distributed residuals (see Chapter 12). A more in-depth assessment of assumptions is found in the multiple regression chapter (Chapter 6).

4.5. Measuring Model Fit with R^2

The residuals are useful for creating a measure of the 'goodness of fit' of a model—how well a model 'fits' the observations overall. A well-fitting model will have small residuals.

Consider an alternative model of response durations that ignores word frequency. The word frequency slope is 0 in this case; the same response duration is predicted regardless of word frequency (see Figure 4.5a). Notice that the residuals shown in

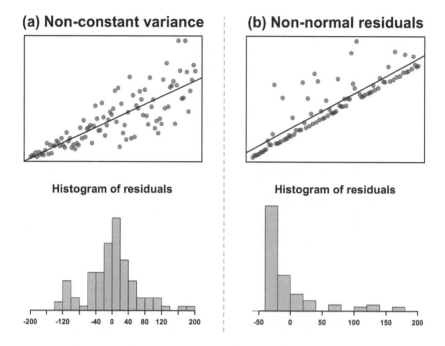

(a) Non-constant variance

(b) Non-normal residuals

Histogram of residuals

Histogram of residuals

Figure 4.4. Clear violations of (a) constant variance and (b) normality

Figure 4.5a are overall larger than the residuals of the model shown in Figure 4.3a. This suggests that the fit for this zero-slope model is worse.

To get an overall measure of 'misfit', the residuals can be squared and summed.[4] The corresponding measure is called 'sum of squared errors' (*SSE*). The regression model with the slope of –70.24 has an *SSE* of 42,609. The zero-slope model has an *SSE* of 152,767, a much larger number. Imagine a whole series of possible slope values, as if you are slowly tilting the line in Figure 4.1. Figure 4.6 shows the *SSE* as a function of different slope values. This graph demonstrates that for the regression model of the response duration data, the estimated slope ($b_1 = -70.28$) results in the smallest residuals. Regression is sometimes called 'least squares regression' because it yields the coefficients that minimize the squared residuals. As the researcher, you only have to supply the equation *response duration=b_0+b_1 * frequency*. In that case, you can think of b_0 and b_1 as 'placeholders' that are to be filled with actual numerical values based on the data.

Let's talk a bit more about the zero-slope model (Figure 4.5a). You can fit such a model by dropping the frequency predictor from your equation, which is the same as assuming that the slope is 0.

$$response\ duration=b_0 \tag{E4.8}$$

---

4 Why squaring? One reason is that this way you get rid of the negative signs. Otherwise the positive and negative residuals would cancel each other out.

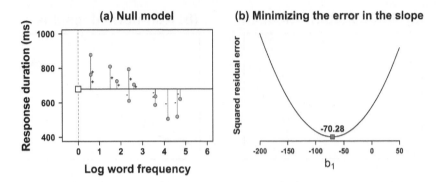

Figure 4.5. (a) A regression model predicting the same response duration regardless of frequency—vertical distances (residuals) are larger than in Figure 4.3a; (b) squared residual error for a whole range of slope estimates—the residuals are minimized for a slope of −70.28

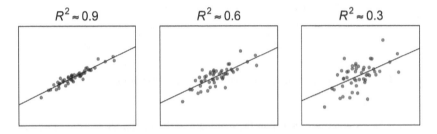

Figure 4.6. Larger residuals yield smaller R^2 values

   This 'intercept-only model' model has not been instructed to look for an effect of frequency. What is the best estimate for the intercept in this case? In the absence of any other information, the best estimate is the mean, as it is the value that is closest to all data points (see Chapter 3).
   The intercept-only model can be a useful a reference model or 'null model' for comparing SSE values. Remember that the main regression model (with frequency predictor) had an SSE of 42,609. Without context, this number is pretty meaningless; it is an unstandardized statistic that changes depending on the metric of the response. For example, if you measured response durations in seconds, rather than milliseconds, the SSE of our main model would shrink from 42,609 to 0.042609. The 'null model' without a slope can be used to put the SSE of the main model into perspective. It can be used to compute a 'standardized' measure of model fit, namely, R^2 ('R squared'). The formula for R^2 is as follows:

$$R^2 = 1 - \frac{SSE_{model}}{SSE_{null}}$$

(E4.9)

The *SSE* achieves this standardization by dividing the main model's *SSE* by the corresponding null model's *SSE*. Division gets rid of the metric. Let's test this with the *SSEs* for both of the models just considered. The *SSE* of the main model (42,609) divided by the *SSE* of the null model (152,767) yields 0.28. The R^2 value then is $1 - 0.28 = 0.72$.

This number can be conceptualized as how much variance is 'described' by a model.[5] In this case, 72% of the variation in response durations can be accounted for by incorporating word frequency into the model. Conversely, 32% of the variation in response durations is due to chance, or due to factors the model omits. In actual linguistic data, R^2 values as high as 0.72 are almost unheard of. Language is complex and humans are messy, so our models rarely account for that much variance.

R^2 is actually a measure of 'effect size'. Specifically, R^2 measures the strength of the relationship between two variables (see Nakagawa & Cuthill, 2007). R^2 values range from 0 to 1. Values closer to one indicate better model fits as well as stronger effects, as shown in Figure 4.6.

Standardized metrics of effect size such as R^2 should always be supplemented by a thorough interpretation of the unstandardized coefficients. You have already done this when computing response durations for different frequency values. When looking at what your model predicts, it is important to use your domain knowledge. As the expert of the phenomenon that you study, you are the ultimate judge about what is a strong or a weak effect.

4.6. A Simple Linear Model in R

As always, you need to load the tidyverse package if you haven't done so already:

```
library(tidyverse)
```

Let's start by generating some random data for a regression analysis, specifically, 50 random normally distributed numbers (see Chapter 3).

```
# Generate 50 random numbers:

x <- rnorm(50)
```

Check the resulting vector (remember that you will have different numbers).

```
# Check first 6 values:

head(x)
```

```
[1] 1.3709584 -0.5646982 0.3631284 0.6328626
[5] 0.4042683 -0.1061245
```

---

5 People often say that R^2 measures the 'variance explained'. An informative blog post by Jan Vanhove (accessed October 16, 2018) recommends against this terminology, among other reasons because it sounds too causal: https://janhove.github.io/analysis/2016/04/22/r-squared

To be able to do anything bivariate, you need *y*-values to go with the *x*-values. Let's say that the intercept is 10 and the slope is 3:

```
# Create y's with intercept = 10 and slope = 3:

y <- 10 + 3 * x
```

Plotting *y* against *x* in a scatterplot (using the optional argument pch = 19 to change the point characters to filled circles) reveals a straight line with no scatter (Figure 4.7, left plot). In other words, *y* is a perfect function of *x* —something that would never happen in linguistic data.

```
plot(x, y, pch = 19)
```

To add noise, the rnorm() function is used a second time to generate residuals.

```
error <- rnorm(50)
y <- 10 + 3 * x + error
```

I want you to notice the similarity between this command and the regression equation ($y = b_0 + b_1 * x + e$) . The error term is what's going to create residuals in the following regression analysis. Next, rerun the plotting command, which yields the plot to the right of Figure 4.7.

```
plot(x., y, pch = 19)
```

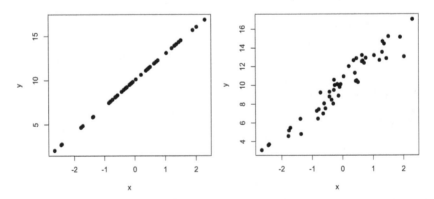

Figure 4.7. Randomly generated data where *y* is a perfect function of *x* (left plot); adding noise to *y* yields the plot on the right

The data is in place. Specifically, you generated random data where *y* depends on *x*. Since you generated the data yourself, you know that the intercept is 10 and the slope is 3. Let's see whether regression is able to retrieve these coefficients from the data.

For this, use the lm() function, which stands for linear model. The syntax used in 'y ~ x' is called 'formula notation', and it can be paraphrased as 'y as a function of x'. Usually, you want to save linear models in R objects so that you can access them for later use.

```
xmdl <- lm(y ~ x)

xmdl

Call:
lm(formula = y ~ x)

Coefficients:
(Intercept)        x
     10.094    2.808
```

R spits out the coefficients, the main outcome of any regression analysis. Notice how these coefficients are fairly close to what was specified when you generated the data (you will see slightly different values). The deviations from the exact values 10 and 3 are not due to the model being wrong (regression always provides the coefficients that minimize the residuals), but due to the random noise that was added to the data.

The fitted() and residuals() functions can be used to retrieve the model's fitted values and residuals. The following command uses the head() function merely to reduce the output to the first six values.

```
head(fitted(xmdl))

        1         2         3         4         5
13.943480  8.508189 11.113510 11.870920 11.229031
        6
 9.795856
```

```
head(residuals(xmdl))

         1          2          3          4
0.49132013 -0.98612217 1.55160234 0.67056756
         5          6
0.07353501  0.16232124
```

The first element of the output of fitted(xmdl) is the prediction for the first data point, and so on. Similarly, the first element of the output of residuals(xmdl) is the deviation of the model's prediction for the first data point, and so on.

A very useful function to apply to linear model objects is summary().

```
summary(xmdl)
```

```
Call:
lm(formula = y ~ x)

Residuals:
  Min    1Q     Median    3Q     Max
-2.6994 -0.6110 0.1832 0.6013 1.5516

Coefficients:
            Estimate Std. Error t value Pr(>|t|)
(Intercept) 10.0939    0.1284    78.61   <2e-16 ***
x            2.8080    0.1126    24.94   <2e-16 ***
---
Signif. codes: 0 '***' 0.001 '**' 0.01 '*' 0.05 '.' 0.1 ' ' 1

Residual standard error: 0.9075 on 48 degrees of freedom
Multiple R-squared: 0.9284,    Adjusted R-squared: 0.9269
F-statistic:  622 on 1 and 48 DF, p-value: < 2.2e-16
```

This function spits out what's called a coefficient table (under the header 'Coefficients'. It also includes useful summary statistics for the overall model fit. In this case, the R^2 value is 0.9284, indicating that the model describes about 93% of the variance in y. Chapter 6 will discuss 'Adjusted R-squared'. The other values (p-value, standard error, etc.) will be discussed in the chapters on inferential statistics (Chapters 9–11).

The coef() function retrieves the coefficients of a linear model. The output of this is a vector of coefficients.

```
coef(xmdl)
```

```
(Intercept)        x
 10.093852   2.807983
```

This vector can be indexed as usual. The following commands retrieve the intercept and slope, respectively.

```
coef(xmdl)[1]
```

```
(Intercept)
   10.09385
```

```
coef(xmdl)[2]
```

```
       x
2.807983
```

```
# Alternative way of indexing (by name):
```

```
coef(xmdl)['(Intercept)']
```

```
(Intercept)
   10.09385
```

```
coef(xmdl)['x']
```

```
       x
2.807983
```

The following command computes the fitted value for an *x*-value of 10:

```
coef(xmdl)['(Intercept)'] + coef(xmdl)['x'] * 10
```

```
(Intercept)
   38.17369
```

Don't be confused by the fact that it says '(Intercept)' at the top of the predicted value. This is just because when R performs a mathematical operation on two named vectors (one being named '(Intercept)', the other one being named 'x'), it maintains the name of the first vector.

The predict() function is useful for generating predictions for fitted models, thus saving you some arithmetic. The function takes two inputs: first, a model which forms the basis for generating predictions; second, a set of new values to generate predictions for.

Let's generate predictions for a sequence of numbers from –3 to +3 using the seq() function. This particular sequence is specified to increase in 0.1 intervals (by = 0.1).

```
xvals <- seq(from = -3, to = 3, by = 0.1)
```

The predict() function needs a data frame or tibble as input. Importantly, the column has to be named 'x', because this is the name of the predictor in xmdl.

```
mypreds <- tibble(x = xvals)
```

Now that you have a tibble to generate predictions for, you can use predict(). The following code stores the output of the mypreds tibble in a new column, called fit.

```
mypreds$fit <- predict(xmdl, newdata = mypreds)
```

```
mypreds
```

```
# A tibble: 50 x 2
      x    fit
```

```
        <dbl>  <dbl>
 1   -2.66    2.63
 2   -2.56    2.92
 3   -2.46    3.20
 4   -2.36    3.48
 5   -2.26    3.76
 6   -2.16    4.04
 7   -2.06    4.32
 8   -1.96    4.60
 9   -1.86    4.88
10   -1.76    5.16
# ... with 40 more rows
```

So, for an *x*-value of –2.66, the model predicts a *y*-value of 2.63, and so on.

4.7. Linear Models with Tidyverse Functions

Let's learn how to do a linear model analysis with tidyverse functions. First, let's put the data into a tibble, then, refit the model.

```
mydf <- tibble(x, y)

xmdl <- lm(y ~ x, data = mydf)
```

The `broom` package (Robinson, 2017) provides tidy model outputs, implemented via the `tidy()` function.

```
library(broom)

# Print tidy coefficient table to console:

tidy(xmdl)
```

```
          term   estimate  std.error  statistic       p.value
1  (Intercept)  10.093852  0.1283994   78.61294  2.218721e-52
2            x   2.807983  0.1125854   24.94091  3.960627e-29
```

The advantage of using `tidy()` is that the output has the structure of a data frame, which means that you can easily index the relevant columns, such as the coefficient estimates.

```
# Extract estimate column from coefficient table:

tidy(xmdl)$estimate
```

```
[1] 10.093852 2.807983
```

The corresponding `glance()` function from the `broom` package gives you a 'glance' of the overall model performance (so far you only know R^2—some of the other quantities will be explained in later chapters).

```
# Check overall model performance:

glance(xmdl)
```

```
 r.squared adj.r.squared      sigma  statistic    p.value
1 0.9283634      0.926871  0.9074764 622.0489 3.960627e-29
  df   logLik        AIC       BIC deviance df.residual
1  2 -65.07199  136.144 141.88 39.52864       48
```

To plot the model with `ggplot2`, you can use `geom_smooth()`. This 'smoother' plots a linear model when specifying the argument `method = 'lm'`. The `geom_point()` function and the `geom_smooth()` function of the following code snippet know what columns to draw the *x*- and *y*-values from because the aesthetic mappings have already been specified in the `ggplot()` command. These two geoms thus share the same set of aesthetic mappings.

```
mydf %>% ggplot(aes(x = x, y = y)) +
  geom_point() + geom_smooth(method = 'lm') +
  theme_minimal()
```

The resultant plot will look similar to Figure 4.9 (plot to the right) with a superimposed regression line. In addition, a gray ribbon is displayed around the regression line, which is the '95% confidence region', which will be explained later (Chapters 9–11).

4.8. Model Formula Notation: Intercept Placeholders

It's important to learn early on that the following two function calls yield equivalent results.

```
xmdl <- lm(y ~ x, data = mydf)

# Same as:

xmdl <- lm(y ~ 1 + x, data = mydf)
```

The intercept is represented by the placeholder '1'.[6] In R's model formula syntax, the intercept is always fitted by default, even if this is not explicitly specified. In other words,

---

6 It is no coincidence that the placeholder is the number one. However, since we don't focus on the mathematical details here, it is beyond the scope of this book to explain why. (Just to pique your interest: this has to do with the matrix algebra machinery that is underlying linear models.)

the shorthand notation 'y ~ x' actually fits the model corresponding to the formula 'y ~ 1 + x'. The second function call is more explicit: it can be paraphrased as 'estimate not only a slope for x, but also an intercept'.

This knowledge can be used to specify an intercept-only model (as was discussed above, see equation E4.8). Let's do this for the data at hand:

```
# Fitting an intercept-only model:

xmdl_null <- lm(y ~ 1, data = mydf)
```

What does this model predict?

```
coef(xmdl_null)
```

```
(Intercept)
   9.993686
```

The model predicts only one number, the intercept. No matter what value x assumes, the model predicts the same number, the mean. Let's verify that the coefficient of this null model is actually the mean of the response:

```
mean(y)
```

```
[1] 9.993686
```

Thus, we have used the intercept notation to fit a linear model that estimates the mean. For now, this will seem rather dull. (Why would you want to do this?) For the later parts of this book, it's important to remember the intercept is implemented by a '1' placeholder.

4.9. Chapter Conclusions

In this chapter, you performed your first regression analysis, regressing response duration on word frequency. The regression line represents a model of the data, specifically a conditional mean. This line is fully specified in terms of an intercept and a slope. The coefficients are the principal outcome of any regression analysis, and they allow making predictions, which are called fitted values. The extent by which the observed data points deviate from the fitted values are called residuals. The residuals are assumed to be normally distributed and homoscedastic (constant variance). For a given model specification, regression minimizes the size of the residuals. A comparison of a model's residuals against a null model's residuals yields a standardized measure of model fit called R^2. Throughout all of this, I emphasized that most of your time should be spend on interpreting the regression coefficients. When doing this, you have to use your field-specific scientific judgment. What does this slope *mean* with respect to your hypothesis?

4.10. Exercises

4.10.1. Exercise 1: Fit the Frequency Model

In this exercise, you will perform the analysis corresponding to Figure 4.1 above. Load in the dataset 'ELP_frequency.csv'. Use mutate() to apply the log10() function to the frequency column (logarithms will be explained in Chapter 5). Fit a model in which response durations are modeled as a function of log frequencies. Create a plot for the relationship between these two variables.

Additional exercise: can you add a horizontal line showing the mean response duration using geom_hline() and the yintercept aesthetic?

4.10.2. Exercise 2: Calculating R^2 by Hand

Run the following lines in R (this requires that you still have the random x-values and random y-values generated in the chapter). Try to make sense of each command. Compare the resulting number to the R^2 value reported by summary(xmdl) or glance(xmdl).

```
xmdl <- lm(y ~ x)
xmdl_null <- lm(y ~ 1)

res <- residuals(xmdl)
res_null <- residuals(xmdl_null)
sum(res ^ 2)
sum(res_null ^ 2)
1 - (sum(res ^ 2) / sum(res_null ^ 2))
```

5   Correlation, Linear, and Nonlinear Transformations

5.1. Centering

A linear transformation involves addition, subtraction, multiplication, or division with a constant value. For example, if you add 1 to the numbers 2, 4, and 6, the resulting numbers (3, 5, and 7) are a linear transformation of the original numbers. Having added a constant value to all numbers has not changed the relations between the numbers, because each number is affected the same way.

Linear transformations are useful, because they allow you to represent your data in a metric that is suitable to you and your audience. For example, consider the response time data from Chapter 4. In this analysis, response durations were measured on a scale of milliseconds. If you wanted to talk about seconds instead, simply divide each response duration by 1000. This won't affect the theoretical conclusions you base on your regression model.

'Centering' is a particularly common linear transformation. This linear transformation is frequently applied to continuous predictor variables. You will learn soon, especially in the chapter on interactions (Chapter 8), how centering facilitates the interpretation of regression coefficients (see Schielzeth, 2010). To center a predictor variable, subtract the mean of that predictor variable from each data point. As a result, each data point is expressed in terms of how much it is above the mean (positive score) or below the mean (negative score). Thus, subtracting the mean out of the variable expresses each data point as a mean-deviation score. The value zero now has a new meaning for this variable: it is at the 'center' of the variable's distribution, namely, the mean.[1]

Centering has immediate effects on the intercept. Recall that the intercept is defined as the y-value for $x = 0$. Have a look at Figure 5.1a, where the intercept is the fitted value for a log frequency of 0. Centering a distribution changes the meaning of 0. If you center the x predictor, the intercept becomes the predicted y-value for the mean of x, as shown in Figure 5.1b. The change from Figure 5.1a to Figure 5.1b should make it clear why 'centering' is called this way, as the intercept is literally drawn to the center

---

[1] Side note: notice that the sample mean is itself an estimate that changes with different samples. As a result of this, your data will be centered in a different way for different samples. It is good for you to know that you can also center using values other than the mean. For example, you could use the middle of a rating scale for centering rating data.

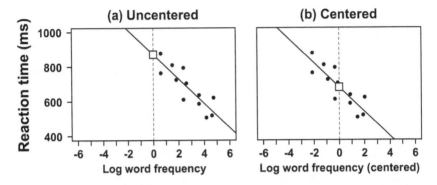

Figure 5.1. Response durations as a function of uncentered and centered word frequencies; intercepts are represented by white squares

of mass of the data. Notice that the slope of the regression line does not change when moving from Figure 5.1a to Figure 5.1b.

Two example coefficient outputs are shown below. Notice how changing the model from an uncentered (Log10Freq) to a centered log frequency predictor (Log10Freq_c) changes the intercept but leaves the slope intact.[2]

```
# uncentered (original):
             Estimate
(Intercept)   870.91
Log10Freq     -70.28

# centered:
             Estimate
(Intercept)   679.92
Log10Freq_c   -70.28
```

In some cases, uncentered intercepts outright make no sense. For example, when performance in a sports game is modeled as a function of height, the intercept is the predicted performance someone of 0 height. After centering, the intercept becomes the predicted performance for a person of average height, a much more meaningful quantity. Thus, centering your predictors affords some interpretational advantages. When dealing with interactions (see Chapter 8), centering is absolutely essential for avoiding mistakes in interpreting a model.

5.2. Standardizing

A second common linear transformation is 'standardizing' or 'z–scoring'. For standardizing, the centered variable is divided by the standard deviation of the sample.

---

2 I use variable names that end with '_c' for all my centered variables.

For example, consider the following response durations from a psycholinguistic experiment:

460ms   480ms   500ms   520ms   540ms

The mean of these five numbers is 500ms. Centering these numbers results in the following:

−40ms   − 20ms   0ms   +20ms   +40ms

The standard deviation for these numbers is ~32ms. To 'standardize', divide the centered data by the standard deviation. For example, the first point, –40ms, divided by 32ms, yields –1.3. Since each data point is divided by the same number, this change qualifies as a linear transformation.

As a result of standardization, you get the following numbers (rounded to one digit):

$-1.3z$ $- 0.6z$ $0z$ $+ 0.6z$ $+1.3z$

The raw response duration 460ms is –40ms (centered), which corresponds to being 1.3 standard deviations below the mean. Thus, standardization involves re-expressing the data in terms of how many standard deviations they are away from the mean. The resultant numbers are in 'standard units', often represented by letter z. Figure 5.2 below shows a distribution that was already discussed in Chapter 3, the emotional valence distribution from Warriner et al. (2013). The figure displays two additional x-axes, one for the centered data, and one for the standardized data. This highlights how linear transformations just change the units on the x-axis; they leave the shape of the data intact.

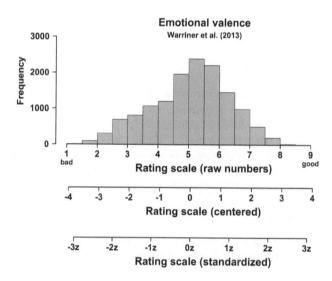

Figure 5.2. Frequency histogram for emotional valence scores with the raw emotional valence scores, the centered emotional valence scores, and the standardized scale

Let us have a look at the output of the response duration model when the log frequency predictor is standardized (`Log10Freq_z`).

```
               Estimate
(Intercept)      679.92
Log10Freq_z     -101.19
```

The slope has changed from −70 (unstandardized) to −101 (standardized). This change is deceiving, however, as the underlying model hasn't changed at all, only in which units the slope is represented. The new slope indicates how much response durations change for 1 standard deviation of the log frequency variable, rather than 1 log frequency value.

So what's standardizing good for? Standardizing is a way of getting rid of a variable's metric. In a situation with multiple variables, each variable may have a different standard deviation, but by dividing each variable by the respective standard deviation, it is possible to convert all variables into a scale of standard units. This sometimes may help in making variables comparable, for example, when assessing the relative impact of multiple predictors (see Chapter 6).

5.3. Correlation

So far, this chapter has only discussed cases in which the predictor is standardized. What if you standardized the response variable as well? In that case, neither the x-values nor the y-values preserve their original metric; both are standardized. Thus, the corresponding regression model will estimate how much change in y standard units results from a corresponding change in x standard units. The resulting slope actually has a special name, it is Pearson's r, the 'correlation coefficient'.

Pearson's r is a standardized metric of how much two variables are correlated with each other. If y increases as x increases, then the correlation coefficient is positive (e.g., age and vocabulary size). If y decreases as x increases, then the correlation coefficient is negative (e.g., frequency and response duration). Correlation coefficients range between −1 and +1. The farther away the coefficient is from zero, the stronger the correlation. Figure 5.3 shows a range of randomly generated datasets with their corresponding correlation coefficients. Compare the scatterplots to the r-values to gauge your intuition about Pearson's r.

Pearson's r being a standardized measure of correlation means that you do not need to know the underlying metric of the data in order to understand whether two variables are strongly correlated or not. Imagine listening to a talk from a completely different field—say, quantum chemistry—and somebody reports that two quantities have a correlation of $r = 0.8$. Then, without knowing anything about quantum chemistry, you can draw a mental picture of what the correlation looks like, in line with the examples seen in Figure 5.3. That said, whether $r = 0.8$ is considered to be a 'high' or 'low' correlation still depends on domain knowledge. For a psychologist or a linguist, this is a really high correlation (our study systems are usually never this well-behaved), for a quantum chemist an r of 0.8 may be low.

You have already seen another standardized statistic in this book, namely, R^2, the 'coefficient of determination' (see Chapter 4). Recall that this number measures the

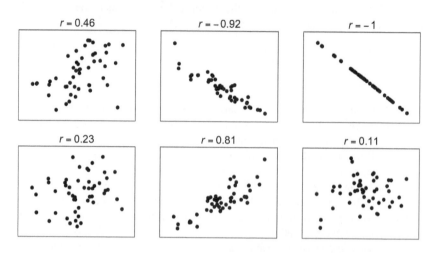

Figure 5.3. Example scatterplots with their corresponding correlation coefficients

'variance described' by a model. When a correlation is very strong, this also means that a lot of variance is described. For simple linear regression models with only one predictor, R^2 is actually the squared correlation coefficient.

5.4. Using Logarithms to Describe Magnitudes

Nonlinear transformations have much less innocuous effects on the data than linear transformations; they do affect the relations between data points. Figure 5.4a shows a distribution of response durations from a psycholinguistic study conducted by Winter and Bergen (2012) (pooled across experiments 1 and 2). Whereas the normal distribution is symmetrical, this distribution is not. The distribution exhibits what is called 'positive skew'. The term 'skew' describes whether a distribution has extreme values in the direction of positive infinity or negative infinity. For the distribution in Figure 5.4a, the skew is 'positive' rather than 'negative', because there are some very large extreme values. The bulk of data actually lies towards smaller values (shorter response durations).

Positive skew is ubiquitous in linguistic data. For example, reaction time data will almost always be skewed, because there is a natural lower limit to how quickly somebody can respond. This limit is determined by how quickly the brain can recognize a stimulus and initiate a motor response, as well as by how quickly it is physically possible to move one's hand to press a button. As a result of these factors, very short response durations are impossible. However, very long durations are possible and they occasionally occur. Many distributions that have a natural lower bound exhibit positive skew.[3]

You may want to subject the response durations seen in Figure 5.4a to a nonlinear transformation before computing means or conducting a regression analysis.

---

3 Besides having a lower bound, there are other, more theoretically interesting reasons for positive skew in linguistic and nonlinguistic data (see Kello, Anderson, Holden, & Van Orden, 2008; Kello, Brown, Ferrer-i-Cancho, Holden, Linkenkaer-Hansen, Rhodes, & Van Orden, 2010).

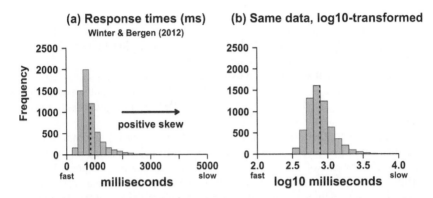

Figure 5.4. The same set of reaction times from Winter and Bergen (2012) presented (a) in raw milliseconds and (b) on a logarithmic scale; the dotted lines indicate means

Chapter 4 talked about the normality and constant variance assumption of regression. If the response variable is very skewed, it is often the case that the residuals of the corresponding models violate one of these assumptions. Thus, performing a nonlinear transformation that gets rid of the skew may make your regression model adhere more strongly to these assumptions. In addition, it is possible that very large values have an undue amount of influence on your regression models (see Baayen, 2008: 92), since they can draw the coefficient estimates away from the bulk of the data.[4]

Probably the most commonly applied nonlinear transformation is the logarithm. The logarithm is best discussed together with exponentiation (taking something to the power of another number). The logarithm takes large numbers and shrinks them. The exponential function takes small numbers and grows them. The logarithm and the exponential function are each other's 'inverses', which is a mathematical term for two functions that reverse each other's effects.

Have a look at the following progression of logarithmically transformed values, which are displayed next to the corresponding exponentiations.

Logarithms	Exponentiation
$\log_{10}(1) = 0$	$10^0 = 1$
$\log_{10}(10) = 1$	$10^1 = 10$
$\log_{10}(100) = 2$	$10^2 = 100$
$\log_{10}(1000) = 3$	$10^3 = 1000$
$\log_{10}(10000) = 4$	$10^4 = 10000$

4 You may find that response durations are often not quite Gaussian even after log-transformation. There are other distributions that can be used in more advanced modeling applications that may provide a better fit to positively skewed data, such as the gamma distribution (see Baayen & Milin, 2010).

This progression shows that the $\log_{10}$ function tracks the order of magnitude of a number. In other words, for these numbers, the $\log_{10}$ counts how many zeros there are. This corresponds to the power value used in exponentiation. For example, taking 10 to the power of 2 (10 x 10) yields 100. Taking the $\log_{10}$ function of 100 extracts the power term, which yields 2 (the number of zeros).

The logarithms in this example move in a linear progression (step-size of one) through the sequence 0–1–2–3–4. On the other hand, the raw numbers progress in much bigger jumps, and the bigger the logarithms, the bigger the jumps between consecutive logarithms.

This also means that large numbers, such as 1000, shrink disproportionately more than small numbers when subjected to a log-transform. The $\log_{10}$ of 10 is 1, which is a difference of 9 between the logarithm and the raw number. The $\log_{10}$ of 1000 is 3, which is a difference of 998. Logarithms thus have a 'compressing' effect on the data, which also means that they change the shape of the distribution, as seen in Figure 5.4b. Because the log-transform affects some numbers more than others, it is not a linear transformation.

The following commands show how to perform the $\log_{10}$ function in R.

```
log10(100)
```

```
[1] 2
```

```
log10(1000)
```

```
[1] 3
```

To get raw scores back from data that has been $\log_{10}$-transformed, you simply need to exponentiate 10 by the corresponding logarithm (taking 10 to the power of 2). This 'back-transforms' the logarithms to the original (uncompressed) scale.

```
10 ^ 2
```

```
[1] 100
```

```
10 ^ 3
```

```
[1] 1000
```

In R, the default log function is `log()`. This is the 'natural logarithm', which is the logarithm to the base e, the special number 2.718282. This number is the default log because e is useful for a number of calculus applications, and this function also features prominently in many areas of statistics (see Chapter 12 and 13). If somebody tells you that they log-transformed the data without stating the base of the logarithm, your best bet is that they used the natural logarithm.[5] One has to

---

5 The base 2 logarithm ($\log_2$) is also quite common, especially in information theory and computer science.

recognize that different logarithms have different compressing effects on the data. In particular, the $\log_{10}$ function compresses the data more strongly than the log function (natural log).

Throughout this book and in my research, I commonly use the $\log_{10}$ function primarily because it is easier to interpret: it's easier to multiply 10s with each other than lots of *e*s. For example, if somebody reports a $\log_{10}$ value of 6.5, then I know that the original number was somewhere between a number that had 6 zeros (one million: 1000000) and 7 zeros (ten million: 10000000). Similar mental calculations aren't as easy for *e*.

Response durations are frequently log-transformed with the natural logarithm, although there are no fixed standards for this (Cleveland, 1984; Osborne, 2005). To demonstrate how to perform calculations with log response durations, let's create a few hypothetical response duration values.

```
RTs <- c(600, 650, 700, 1000, 4000)

RTs
```

```
[1] 600 650 700 1000 4000
```

Let's assume that these numbers are measured in milliseconds (4000 = 4 seconds, and so on). Next, generate a log-transformed version of this vector.

```
logRTs <- log(RTs)
```

Let's check the results.

```
logRTs
```

```
[1] 6.396930 6.476972 6.551080 6.907755 8.294050
```

The resulting numbers are much smaller than the raw response durations, and the largest response duration (4000ms) is now much closer to the other numbers. Crucially, these numbers now represent a fundamentally different quantity, namely, magnitudes of response times, rather than response times.

What if you wanted to 'undo' the log-transform? For this, you have to use the inverse of the natural logarithm, which is implemented in R via the exponential function $\exp()$.[6]

```
exp(logRTs)   # undo the logging
```

```
[1] 600 650 700 1000 4000
```

---

6 For the $\log_{10}$ function, there is no equivalent of the $\exp()$ function. Instead, you have to use powers of 10. If you had a $\log_{10}$ response duration of 3.5, the following command gives you the corresponding raw number: 10 ^ 3.5

As mentioned earlier, the logarithm has many properties that are useful in regression modeling (such as the ability to make the residuals more 'well-behaved'). However, you have to continuously remind yourself that when working with a logarithmically transformed variables, you are fundamentally changing the nature of the data. On this point, it's worth noting that many cognitive and linguistic phenomena are scaled logarithmically. To illustrate this, do you share the intuition that the difference between the numbers 1 and 2 *seems* bigger than the difference between the numbers 5,001 and 5,002? Of course, both pairs differ by exactly the same amount, but one difference may appear larger to you than the other, which may result from your mental number line being logarithmically scaled (see Dehaene, 2003). Estimates of the intensity of perceptual stimuli, such as weights and durations, also follow logarithmic patterns (Stevens, 1957). Logarithmic scaling may even have deep roots in our neural architecture (Buzsáki & Mizuseki, 2014).

In linguistics, the seminal work by Zipf (1949) has shown that a number of linguistic variables, such as word length or the number of dictionary meanings, track the logarithm of word frequency, rather than the raw frequency. Smith and Levy (2013) discuss evidence that processing times are scaled logarithmically. Many acoustic variables in phonetics are also scaled logarithmically or semi-logarithmically in perception, such as loudness (the decibel scale) or pitch (the bark scale). Thus, not only do logarithms help researchers fit more appropriate regression models, they are also often theoretically motivated.

5.5. Example: Response Durations and Word Frequency

Now that you know about logarithms, you will be able to perform the full analysis discussed in Chapter 4 yourself. After setting your working directory, load in the 'ELP_frequency.csv' file.

```
library(tidyverse)
library(broom)

ELP <- read_csv('ELP_frequency.csv')

ELP
```

```
# A tibble: 12 x 3
   Word        Freq     RT
   <chr>      <int>  <dbl>
 1 thing      55522   622.
 2 life       40629   520.
 3 door       14895   507.
 4 angel       3992   637.
 5 beer        3850   587.
 6 disgrace     409   705.
 7 kitten       241   611.
 8 bloke        238   794.
```

```
 9 mocha         66   725.
10 gnome         32   810.
11 nihilism       4   764.
12 puffball       4   878.
```

Let's start by creating logarithmically transformed frequency and response time columns using mutate()

```
ELP <- mutate(ELP,
              Log10Freq = log10(Freq),
              LogRT = log(RT))
```

The new column Log10Freq contains the $\log_{10}$ of Freq. I am using the logarithm to the base 10 here for interpretability, but one could have also used the natural logarithm log(), which compresses the data less strongly. The third argument of this mutate() command creates the LogRT column, which contains the log-transformed response durations. Here, I use the natural logarithm, which is most commonly used in psycholinguistics.

Let's check the tibble:

```
ELP
```

```
# A tibble: 12 x 5
    Word      Freq    RT Log10Freq LogRT
    <chr>    <int> <dbl>    <dbl> <dbl>
 1 thing     55522  622.     4.74  6.43
 2 life      40629  520.     4.61  6.25
 3 door      14895  507.     4.17  6.23
 4 angel      3992  637.     3.60  6.46
 5 beer       3850  587.     3.59  6.38
 6 disgrace    409  705      2.61  6.56
 7 kitten      241  611.     2.38  6.42
 8 bloke       238  794.     2.38  6.68
 9 mocha        66  725.     1.82  6.59
10 gnome        32  810.     1.51  6.70
11 nihilism      4  764.     0.602 6.64
12 puffball      4  878.     0.602 6.78
```

For pedagogical purposes, Chapter 4 regressed raw response times on log frequencies. Here, log response times will be regressed on log frequencies. For this particular dataset, it does not make much of a difference (however, taking more data from the English Lexicon Project into account would show that the log-transformed response durations are preferred because they align more strongly with the assumptions of regression).

To illustrate the effects of log-transforming the frequency predictor, it's helpful to plot the response durations against the raw frequencies first (Figure 5.5, left plot). The following code plots the data as text with geom_text(). The regression model is

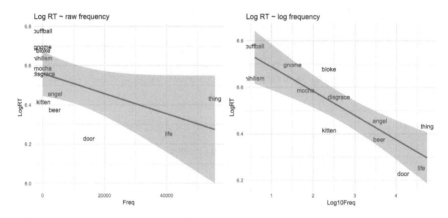

Figure 5.5. Left: log response durations as a function of raw word frequencies with model fit; right: log response durations as a function of log word frequencies

added with `geom_smooth(method = 'lm')` (the meaning of the gray regions will be explained in Chapters 9, 10, and 11). Finally, the `ggtitle()` function adds a title to the plot.

```
ELP %>% ggplot(aes(x = Freq, y = LogRT, label = Word)) +
  geom_text() +
  geom_smooth(method = 'lm') +
  ggtitle('Log RT ~ raw frequency') +
  theme_minimal()
```

The left-hand plot of Figure 5.5 shows a few words to be extremely frequent, such as *life* and *thing*. Moreover, a lot of the words with lower frequencies are scrunched together to the left side of the plot. Figure 5.5 plots the same data against log frequency values, which is achieved by the following code:

```
ELP %>% ggplot(aes(x = Log10Freq, y = LogRT,
                   label = Word)) +
  geom_text() +
  geom_smooth(method = 'lm') +
  ggtitle('Log RT ~ log frequency') +
  theme_minimal()
```

Notice that when frequency is scaled logarithmically, there is a nice linear relationship to response durations. Let's build a model of this.

```
ELP_mdl <- lm(LogRT ~ Log10Freq, data = ELP)
```

Let's have a look at the coefficients via the `tidy()` function from the `broom` package.

```
tidy(ELP_mdl)
```

```
        term   estimate   std.error  statistic      p.value
1 (Intercept) 6.7912813 0.06113258 111.091039 8.564422e-17
2  Log10Freq -0.1042491 0.02006108  -5.196583 4.032836e-04
```

Without performing any mathematical tricks, you can immediately interpret the sign of the coefficient. Notice that the `Log10Freq` coefficient is a negative number, which means that, as log frequencies increase, log response durations become shorter; more frequent words are processed faster. While it is often useful to fit a regression model on log-transformed data (due to assumptions, influential values, etc.), you may still want to *interpret* your model in terms of raw milliseconds. Thus, you may want to 'back-transform' the log predictions of your model into the corresponding millisecond predictions. Let's generate the predicted response duration for two representative word frequencies, 10 and 1000. To make the following code less chunky, let's first extract the coefficients from the model equation.

```
b0 <- tidy(ELP_mdl)$estimate[1]   # intercept

b1 <- tidy(ELP_mdl)$estimate[2]   # slope
```

If you want to generate predictions for words with the frequencies 10 and 1000, what log values do you need to plug into the equation? Here it becomes useful to use base 10 for the frequency predictor. A word with a frequency of 10 corresponds to a $\log_{10}$ of 1 since $10^1 = 10$. A word with a frequency of 1000 corresponds to a $\log_{10}$ of 3 since $10^3 = 10*10*10 = 1000$. So, let's fill out the equation for 1 and 3:

```
logRT_10freq <- b0 + b1 * 1

logRT_1000freq <- b0 + b1 * 3
```

The objects `logRT_10freq` and `logRT_1000freq` now contain the predicted log response durations for the word frequencies 10 and 1000.

```
logRT_10freq
```

```
[1] 6.687032
```

```
logRT_1000freq
```

```
[1] 6.478534
```

For a word frequency of 10, the model predicts a log response duration of about 6.69; for a word frequency of 1000, it predicts about 6.48. To convert these numbers back to raw response durations, you need to apply the inverse of the natural logarithm, the exponential function `exp()`.

```
exp(logRT_10freq)
```

```
[1]  801.9387
```

```
exp(logRT_1000freq)
```

```
[1]  651.0159
```

The model predicts that response durations are about 807ms for words with a frequency of 10. A word that is two magnitudes more frequent (raw frequency = 1000) is predicted to have a response duration of 652ms. That's a difference of about 150ms, which is quite a large difference for a psycholinguistic study.

5.6. Centering and Standardization in R

This next exercise provides additional practice for centering and standardizing. In many ways, the following linear transformations will seem rather 'cosmetic' to you, and you may even wonder at this stage why they are necessary. You can think of the following exercise as a 'dry run' that allows you to rehearse centering and standardizing data in a safe space. The importance of these linear transformations will be felt more strongly in later chapters.

The following command uses `mutate()` to compute all linear transformations in one step. The mean is subtracted to generate the centered variable `Log10Freq_c`. This centered variable is then divided by the standard deviation.

```
ELP <- mutate(ELP,
       Log10Freq_c = Log10Freq - mean(Log10Freq),
       Log10Freq_z = Log10Freq_c / sd(Log10Freq_c))
```

Let's have a look at the different frequency columns next to each other:

```
select(ELP, Freq, Log10Freq, Log10Freq_c, Log10Freq_z)
```

```
# A tibble: 12 x 4
     Freq Log10Freq Log10Freq_c Log10Freq_z
    <int>     <dbl>       <dbl>       <dbl>
1   55522      4.74        2.03        1.41
2   40629      4.61        1.89        1.31
3   14895      4.17        1.46        1.01
4    3992      3.60       0.884       0.614
```

5	3850	3.59	0.868	0.603
6	409	2.61	-0.106	-0.0736
7	241	2.38	-0.336	-0.233
8	238	2.38	-0.341	-0.237
9	66	1.82	-0.898	-0.624
10	32	1.51	-1.21	-0.842
11	4	0.602	-2.12	-1.47
12	4	0.602	-2.12	-1.47

R also has the built-in function `scale()`, which by default standardizes a vector. If you only want to center, you can override the default by specifying `scale = FALSE`. The following command repeats the linear transformations discussed above, but this time using `scale()`.[7]

```
# Same as before, different approach:

ELP <- mutate(ELP,
        Log10Freq_c = scale(Log10Freq, scale = FALSE),
        Log10Freq_z = scale(Log10Freq))
```

With the data in place, you can create linear models with the centered and standardized predictors:

```
ELP_mdl_c <- lm(LogRT ~ Log10Freq_c, ELP)   # centered
ELP_mdl_z <- lm(LogRT ~ Log10Freq_z, ELP)   # z-scored
```

Let's compare the coefficients from the different models:

```
tidy(ELP_mdl) %>% select(term, estimate)
```

```
          term    estimate
1  (Intercept)   6.7912813
2    Log10Freq  -0.1042491
```

```
tidy(ELP_mdl_c) %>% select(term, estimate)
```

```
          term    estimate
1 (Intercept)   6.5079660
2 Log10Freq_c  -0.1042491
```

---

7 Newcomers to statistical modeling seem to like the `scale()` function, but I recommend constructing centered and standardized variables 'by hand'. This is not only more explicit, but `scale()` does a number of things behind the scenes that may sometimes cause problems.

```
tidy(ELP_mdl_z) %>% select(term, estimate)   # slope changes
```

```
            term    estimate
1 (Intercept)    6.507966
2 Log10Freq_z   -0.150103
```

First, compare xmdl to xmdl_c. There is no change in the slope, but the intercept is different in the centered model. In both models, the intercept is the prediction for $x = 0$, but $x = 0$ corresponds to the average frequency in the centered model. Second, compare xmdl_c and xmdl_z. The intercepts are the same because, for both models, the predictor has been centered. However, the slope has changed because a change in one unit is now a change in 1 standard deviation.

To convince yourself that each one of these models is just a different representation of the same underlying relationship, you can look at the overall model statistics with glance()—in which case, you'll find that everything stays the same. For example, each of the models describes exactly the same amount of variance:

```
glance(ELP_mdl)$r.squared
```

```
[1] 0.7183776
```

```
glance(ELP_mdl_c)$r.squared
```

```
[1] 0.7183776
```

```
glance(ELP_mdl_z)$r.squared
```

```
[1] 0.7183776
```

If, however, you compare any of these models to another one that differs in a *nonlinear* transformation, you get a different R^2 value, which indicates a different fit.

```
glance(lm(LogRT ~ Freq, ELP))$r.squared
```

```
[1] 0.2913641
```

Finally, let's compare regression to correlation. You can calculate the correlation coefficient using the cor() function. To teach you some more R, the following command uses the with() function, which makes the ELP tibble available to the cor() function. That way, you don't have to use the dollar sign to index the respective columns, since the function already knows that you are operating on the ELP tibble.

```
with(ELP, cor(Log10Freq, LogRT))
```

```
[1] -0.8542613
```

The result of this command shows that Pearson's correlation coefficient is $r = -0.85$, a very strong negative correlation. The fact that the correlation is negative means that, as frequency increases, response durations become shorter.

For pedagogical reasons, it's useful to recreate the output of cor() using the lm() function. For this, you simply have to run a regression model where both the response and the predictor are z-scored. In addition, you need to prevent the regression model from attempting to estimate an intercept. Notice that the intercept does not have to be estimated, as we know it's zero anyway (since both x and y are standardized). To tell the linear model function that you don't want to estimate an intercept, you add '−1' to the model equation (remember from Chapter 4.8 that '1' acts as a placeholder for the intercept).

```
ELP_cor <- lm(scale(LogRT) ~ -1 + Log10Freq_z, ELP)

tidy(ELP_cor) %>% select(estimate)
```

```
    estimate
1 -0.8542613
```

The slope of this linear model is exactly equal to Pearson's r.

5.7. Terminological Note on the Term 'Normalizing'

A quick terminological aside: the term 'normalizing' is used quite frequently across various disciplines, but, like so many statistical terms, it means different things to different people. Some people use the term 'normalizing' to describe standardization. This is perhaps confusing, because the distribution doesn't look any more or less 'normal' (Gaussian) after standardization. On the other hand, sometimes the term 'normalizing' is used to refer to nonlinear transformations, since it tends to make positively skewed data look more normal.

To make matters worse, phoneticians have yet another use for the term 'normalizing' that's worth discussing here for a second. Particularly when studying speech production, researchers sometimes 'normalize' speaker characteristics out of a dataset. For example, each speaker has a different vocal tract, which affects vowel acoustics. If you wanted vowel acoustics 'unaffected' by a speaker's idiosyncratic physiology, you could take a particular speaker's mean and standard deviation to standardize all of the data points from that speaker. For the next speaker, you would do the same, using their respective mean and standard deviation. This practice ceases to be a linear transformation, because the means and standard deviations used for standardizing are different across speakers. So, this practice *does* affect the relationships between the data points. z-scoring 'within speaker' has little to do with the type of linear transformation that is discussed here, which merely serves to express regression slopes in standard units.

5.8. Chapter Conclusions

In this chapter, you were introduced to linear transformations, of which the two most common are centering and standardizing. Being a linear transformation, centering and standardizing have rather innocuous effects on a model, merely changing its representation. Standardizing will become useful when dealing with models that contain many predictors (Chapter 6), and centering will become a lifesaver when dealing with interactions (Chapter 8).

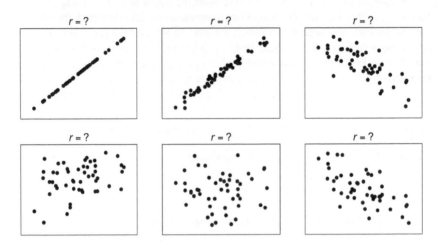

Figure 5.6. 'Guesstimate' the correlation coefficients (Pearson's *r*) for these randomly
generated datasets

In addition, this chapter dealt with nonlinear transformations, which change more
than just model representation. Nonlinear transformations are *not* cosmetic changes.
Instead, they result in fundamentally different models. This chapter focused on the
logarithm, which is commonly used in linguistics to transform response duration data
or word frequency data. The logarithm expresses a phenomenon in orders of magnitudes, and has a compressing effect on the data.

5.9. Exercise

5.9.1. Guessing Correlation Coefficients

It's very important to have a good intuition about Pearson's *r* . Attempt to 'guesstimate' the correlation coefficients for the plots shown in Figure 5.6. It's OK to be off
by quite a bit here—we are not doing rocket science.

6 Multiple Regression

6.1. Regression with More Than One Predictor

The true power of the linear model framework comes to the fore once we move from simple linear regression to multiple regression. That is, we move from models with only one predictor to models with multiple predictors. This is still nothing more than modeling a mean. However, the mean is now conditioned on multiple variables within the same model. In doing so, the right-hand side of the regression equation is expanded to include more than just one slope:

$$y = b_0 + b_1 x + b_2 x + \ldots + e \tag{E6.1}$$

Chapter 4 discussed the analysis of response durations as a function of log word frequency. This analysis only dealt with a small subset of words. The present chapter widens the scope of this analysis to incorporate more words (~33,000 words from the English Lexicon Project; Balota et al., 2007) and more predictors.

Let's start by running a simple linear regression of response durations as a function of log frequencies, which yields the following estimated coefficients for the full dataset:

$$RT = 900ms + (-90) * log\ frequency \tag{E6.2}$$

The predicted average response duration for the word decreases by 90ms for every increase in log frequency by one unit. The slope for the data discussed in Chapter 4 was −70, which was slightly less steep.

The regression model turns out to describe 38% of the variance in response durations ($R^2 = 0.38$). But can all of this variance really be attributed to frequency? Any regression model only knows about what its user tells it to look at. Because this model only knows about frequency, the effect of other variables that are correlated with frequency may be conflated with the frequency effect. For example, it's known since Zipf's work that more frequent words tend to be shorter than less frequent words (Zipf, 1949). Clearly, one can expect that shorter words are read more quickly, so perhaps the frequency effect seen in equation E6.2 is in part due to word length. To assess the influence of frequency while holding word length constant, you can add word length (number of phonemes) as an additional predictor to the model.

$$RT = b_0 + b_1 * log\ frequency + b_2 * word\ length \tag{E6.3}$$

Running the actual regression in R yields the following estimates for b_0, b_1, and b_2.

$$RT = 750ms + (-70)*frequency + 20*length \qquad (E6.4)$$

This model now describes 49% of the variance in response durations ($R^2 = 0.49$). Adding word length has increased the amount of variance described, which suggests that this predictor is capturing something meaningful about reading times. The coefficient of the new word length predictor is positive (+20), which indicates that, as word length increases, response durations increase as well. In other words, longer words take more time to read.

Notice that the intercept has dropped from 900ms in the model without word length to 750ms in the model with word length. To explain this change, it is important to remind yourself of the meaning of the intercept: the intercept is the predicted value when $x = 0$. In this model, 900ms is the prediction for a word with 0 log frequency <u>and</u> 0 word length. As word lengths cannot be 0, this intercept is not particularly meaningful, which is why it might be a good idea to center both predictors (see Chapter 5), in which case the intercept will change to be the response duration for a word with average frequency and average length.

What is most interesting to us right now is that the magnitude of the frequency coefficient has decreased from –90 to –70 as a result of adding the word length predictor, a change of about 20%. This suggests that the effect of frequency in the first model was in fact somewhat confounded with word length. Once one controls for word length, the effect of frequency does not appear as strong.

In the present example, both frequency and word length are shown to influence response durations. However, in some cases, adding another predictor can completely nullify the effects of an existing predictor. For example, in a language acquisition study, you could model vocabulary size as a function of body height. Some thinking will tell you that there has to be a relationship between body height and vocabulary size: vocabularies increase as kids grow older, which happens to coincide with them becoming taller. However, this height effect is almost surely entirely attributable to the concomitant age differences. When vocabulary size is modeled as a function of both body height and age, the height predictor is likely going to have a tiny effect, if at all.

It is important to recognize that multiple regression is not the same as running lots of simple regressions (see Morrissey & Ruxton, 2018 for an illustrative explanation). The coefficients in multiple regression assume a different meaning; specifically, each coefficient becomes a *partial* regression coefficient, measuring the effect of one predictor while holding all other predictors constant.

Let's think about what this means in the case of the word frequency model above. The partial regression coefficient of –70 is to be interpreted as the change in response durations as a function of word frequency, *while holding word length constant*. Come to think of it, this is actually quite a strange quantity, because in actual language word length and word frequency are of course correlated with each other. However, the multiple regression coefficients allow making predictions for highly frequent words that are long, or for very infrequent words that are short. Thus, the true purpose of moving from simple regression to multiple regression is to disentangle the direct effects of specific variables in a study, with each coefficient representing a variable's influence while statistically controlling for the other variables.

6.2. Multiple Regression with Standardized Coefficients

In this section, you will retrace the steps of an actual analysis from Winter et al. (2017). The data was already briefly introduced in Chapter 2. To remind you, the concept of iconicity describes whether the form of words resembles their meaning. For arbitrary words, there is no apparent correspondence between form and meaning; for example, the word *purple* does not sound like the color the word denotes. However, onomatopoeic words such as *beeping*, *buzzing*, and *squealing* mimic the respective sounds.

In Winter et al. (2017), we measured iconicity via a rating scale, asking native speakers a simple question: 'How much does this word sound like what it means?' The scale went from –5 (the word sounds like the opposite of what it means) to +5 (the word sounds exactly like what it means). There are several shortcomings with an approach that relies on native speaker intuitions (see discussion in Perry, Perlman & Lupyan, 2015), but for now we are going to accept the fact that the iconicity ratings give us as a rough measure of a word's iconicity. In the following analysis, you will assess which factors predict iconicity ratings.

First, load the required packages (`tidyverse` and `broom`) and data into your current R session:

```
library(tidyverse)
library(broom)

icon <- read_csv('perry_winter_2017_iconicity.csv')

icon %>% print(n = 4, width = Inf)
```

```
# A tibble: 3,001 x 8
  Word    POS              SER  CorteseImag   Conc  Syst
  <chr>   <chr>            <dbl>       <dbl>  <dbl> <dbl>
1 a       Grammatical NA                 NA   1.46   NA
2 abide   Verb        NA                 NA   1.68   NA
3 able    Adjective   1.73               NA   2.38   NA
4 about   Grammatical 1.2                NA   1.77   NA
      Freq Iconicity
     <int>     <dbl>
1 1041179     0.462
2     138      0.25
3    8155     0.467
4  185206      -0.1
# ... with 2,997 more rows
```

In our 2017 paper (Winter et al., 2017), we modelled iconicity as a function of sensory experience (SER), log word frequency, imageability, and what is called 'systematicity' (see Monaghan, Shillcock, Christiansen, & Kirby, 2014; Dingemanse, Blasi,

Lupyan, Christiansen, & Monaghan, 2015). I won't go into the details of each of these variables, but you should know that they are all continuous variables.[1] Moreover, it's important to mention that each one was included in our study because we had motivated hypotheses for doing so (check out the discussion in Winter et al., 2017). It's a good idea to have a clear plan before you conduct a regression analysis.

Before fitting the regression model, we should log-transform the frequency predictor (see Chapter 5).

```
icon <- mutate(icon, Log10Freq = log10(Freq))
```

Now that we have all variables in place, let's fit a multiple regression model. Simply list all predictors separated by plus signs. The order in which you enter the predictors into the formula does not matter, as everything is estimated simultaneously.

```
icon_mdl <- lm(Iconicity ~ SER + CorteseImag +
            Syst + Log10Freq, data = icon)
```

How much variance in iconicity ratings is accounted for by the entire model?

```
glance(icon_mdl)$r.squared
```

```
[1] 0.2124559
```

The model accounts for about 21% of the variance in iconicity ratings (I'd say that's quite high for linguistic data). Next, investigate the model coefficients.

```
tidy(icon_mdl) %>% select(term, estimate)
```

```
           term      estimate
1 (Intercept)    1.5447582
2          SER    0.4971256
3 CorteseImag   -0.2632799
4         Syst  401.5243106
5    Log10Freq   -0.2516259
```

To facilitate the following discussion, it makes sense to round the coefficients. For this, the round() function can be used. This function takes two arguments: first, a vector of numbers; second, a number indicating to how many decimals the vector should be rounded to. For the code below, the '1' indicates that numbers are rounded to one decimal.[2]

---

[1] The sensory experience measure comes from Juhasz and Yap's (2013) rating study. The word frequency data comes from SUBTLEX (Brysbaert & New, 2009). The imageability data comes from Cortese and Fugett (2004). The systematicity measure comes from Monaghan et al. (2014).

[2] A quick note on rounding: for the write-up of your results, you often have to report rounded numbers, which is what most journals require. You may ask: 'But doesn't rounding involve the loss of

```
tidy(icon_mdl) %>% select(term, estimate) %>%
  mutate(estimate = round(estimate, 1))
```

```
         term estimate
1 (Intercept)      1.5
2         SER      0.5
3 CorteseImag     -0.3
4        Syst    401.5
5    Log10Freq    -0.3
```

To put the output into the format of a predictive equation, you simply need to read off the `estimate` column from top to bottom, lining up the terms from left to right via addition.

$$iconicity = 1.5 + 0.5 * SER + (-0.3) * Imag + 401.5 * Syst$$
$$+ (-0.3) * log\,freq \qquad (E6.5)$$

As always, plugging in some numbers helps to get a feel for this equation. Let's derive the prediction for a word that has a sensory experience rating of 2, an imageability rating of 1, a systematicity of 0, and a log frequency of 0.[3] Then the equation becomes:

$$iconicity = 1.5 + 0.5 * 2 + (-0.3) * 1 + 401.5 * 0 + (-0.3) * 0 \qquad (E6.6)$$

The systematicity and frequency terms drop out through multiplication with 0. Doing the arithmetic yields a predicted iconicity rating of 2.2.

You may think that the `Syst` predictor has the strongest influence on iconicity, as it is associated with the largest coefficient in the coefficient table. However, there's a catch. Let's check the range of the systematicity.[4]

```
range(icon$Syst, na.rm = TRUE)
```

```
[1] -0.000481104 0.000640891
```

---

precision?' Yes, it does, but that's a good thing. You shouldn't invoke a sense of 'false precision'. After all, linguistics isn't rocket science and estimates vary from sample to sample anyway. Displaying numbers with lots of decimals may give the false impression that these decimals actually play a big role. On top of that, nobody wants to read numbers with lots of decimals, which usually detracts from the core message.

3 A log frequency of 0 corresponds to a raw word frequency of 1, since 100=1.

4 Because the systematicity variable has missing values (NA), you need to specify the additional argument na.rm = TRUE. This will result in dropping NA values when computing the range. Notice, furthermore, that the lm() function automatically excludes data points for which there are missing values.

The `Syst` predictor ranges from one really small negative number to an equally small positive number. The regression coefficient of the systematicity predictor (+401.5) is reported in terms of a one-unit change. Given this very narrow range, a one-unit change is a massive jump, in fact, it exceeds the range for which the systematicity measure is attested.

This is a telling example of how you always have to keep the metric of each variable in mind when performing multiple regression. What does a one-unit change mean for frequency? What does a one-unit change mean for imageability? And so on.

Standardization can be used to make the slopes more comparable (see Chapter 5). Remember that standardization involves subtracting the mean (centering) and subsequently dividing the centered scores by the standard deviation. Since each variable has a different standard deviation, this involves dividing each variable by a different number. In effect, you are dividing the metric of the data out of each variable, which makes the variables more comparable. The following code achieves this for all predictors:

```
icon <- mutate(icon,
       SER_z = scale(SER),
       CorteseImag_z = scale(CorteseImag),
       Syst_z = scale(Syst),
       Freq_z = scale(Log10Freq))
```

Now that all variables are standardized, you can fit a new linear model.

```
icon_mdl_z <- lm(Iconicity ~ SER_z + CorteseImag_z +
          Syst_z + Freq_z, data = icon)
```

To reiterate a point made in Chapter 5, let's begin by verifying that the R^2 value hasn't changed.

```
glance(icon_mdl_z)$r.squared
```

```
[1] 0.2124559
```

The fact that the R^2 value is the same as mentioned above is a reminder that the underlying model hasn't changed through standardization. However, the coefficients are represented in different units:[5]

```
tidy(icon_mdl_z) %>% select(term, estimate) %>%
  mutate(estimate = round(estimate, 1))
```

---

5 Let me use this opportunity to sneak in another footnote on rounding. When you want to report rounded numbers in your paper, it's best to do the rounding in R using the `round()` function. Doing the rounding in your head is error-prone. Ideally, your knitted markdown file contains exactly the same (rounded) numerical values that are reported in the write-up of your results.

```
              term estimate
1      (Intercept)      1.3
2           SER_z       0.5
3  CorteseImag_z       -0.4
4          Syst_z       0.0
5          Freq_z      -0.3
```

For these coefficients, a 'one-unit change' always corresponds to a change of 1 standard deviation. Given this rescaling of the predictors, it is now apparent that the sensory experience predictor has the biggest effect on iconicity, with a slope of +0.5. This translates into the following statement: 'For each increase in SER by 1 standard deviation (holding all other variables constant), iconicity increases by +0.5.' Notice that the systematicity predictor is now close to 0.

6.3. Assessing Assumptions

The current iconicity model is ideal for picking up some points that were left over from Chapter 4's discussion of the assumptions of regression (normality of residuals, homoscedasticity of residuals). So, this section will seize the opportunity and go off at a tangent.

It's generally recommended to assess normality and homoscedasticity visually. To assess whether the residuals are normally distributed, one may draw a histogram of the residuals, as shown in Figure 6.1a. For the `icon_mdl` model discussed above, the distribution of the residuals looks good. A better way to graphically explore the normality assumption is via a 'quantile-quantile' plot (Q-Q plot), shown in Figure 6.1b. When the sample quantiles in this plot assemble into a straight line, the residuals conform with the normal distribution.[6]

According to the constant variance assumption, the error should be equal across the fitted values. This can be investigated via a 'residual plot', as shown in Figure 6.1c. This plots the residuals (*y*-axis) against the fitted values (*x*-axis). When the constant variance assumption is satisfied, the spread of the residuals should be approximately equal across the range of fitted values; that is, the residual plot should look basically look like a blob. Any systematic patterns in the residual plot are reasons to be concerned. The residual plot in Figure 6.1c looks pretty good. Perhaps the variance of the residuals funnels out a tad bit towards higher fitted values, but there clearly is no drastic violation of the constant variance assumption.

Newcomers to regression modeling often find it discomforting that the assumptions are assessed visually. In fact, formal tests for checking assumptions do exist, such as the Shapiro-Wilk test of normality. However, applied statisticians

---

6 To create this plot, every residual is transformed into a percentile (or quantile), e.g., the first residual of `icon_mdl` is –1.1, which is the 13.8th percentile (13.8% of the residuals are below this number). The question the Q-Q plot answers is: what is the corresponding numerical value of the 13.8th percentile on the normal distribution? If the values are the same, they will fit on a straight line, which indicates that the two distributions (the distribution of the residuals and the theoretical normal distribution) are very similar.

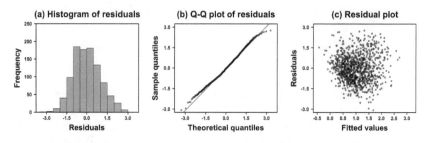

Figure 6.1. (a) Histogram, (b) Q-Q plot, and (c) residual plot of `icon_mdl`

generally prefer visual diagnostics (Quinn & Keough, 2002; Faraway, 2005, 2006: 14; Zuur et al., 2009, Zuur, Ieno, & Elphick, 2010). The most important reason for using graphical validation of assumptions is that it tells you more about your model and the data.[7] For example, the residuals may reveal a hidden nonlinearity, which would suggest adding a nonlinear term to your model (see Chapter 8). Or the residuals may reveal extreme values that are worth inspecting in more detail. One should also remember that a model's adherence to the normality and constant variance assumptions is not a strict either/or. Faraway (2006: 14) says that "It is virtually impossible to verify that a given model is exactly correct. The purpose of the diagnostics is more to check whether the model is not grossly wrong."

Let's implement the graphical diagnostics discussed above. First, extract the residuals of the model into an object called `res`. This is not strictly speaking necessary, but it will save you some typing further down the line.

```
res <- residuals(icon_mdl_z)
```

For the following plots, the base R plotting functions will suit our needs better than `ggplot()`. Let's create a plot matrix with three plots in one row, which is achieved by setting the `mfrow` argument of the `par()` function (the name stands for plotting 'parameters') to `c(1, 3)`, which will create a one-row-by-three-column matrix. The `hist()` function will plot into the first column of this matrix. The `qqnorm()` creates a Q-Q plot. A Q-Q line can be added into the existing plot with `qqline()`. In the final step, a residual plot is added to the third column of the plot matrix. Executing all of the following commands in one sequence will yield a series of plots that looks similar to Figure 6.1.

```
# Set graphical parameters to generate plot matrix:

par(mfrow = c(1, 3))
```

---

7 Here are some other reasons: each of these tests also has assumptions (which may or may not be violated), the tests rely on hard cut-offs such as significance tests (even though adherence to assumptions is a graded notion), and the tests may commit Type I errors (false positives) or Type II errors (false negatives) (see Chapter 10 for an explanation of these concepts).

```
# Plot 1, histogram:

hist(res)

# Plot 2, Q-Q plot:

qqnorm(res)
qqline(res)

# Plot 3, residual plot:

plot(fitted(icon_mdl_z), res)
```

To gauge your intuition about residual plots, the following code (adapted from Far-away, 2005) gives you an idea of what good residual plots should look like. The code uses a `for` loop to repeat a plotting command nine times. As a result, the 3 X 3 plot matrix that is set up by `par()` is iteratively filled with plots of 50 random data points.

```
par(mfrow = c(3, 3))   # setup 3 X 3 plot matrix

for (i in 1:9) plot(rnorm(50), rnorm(50))
```

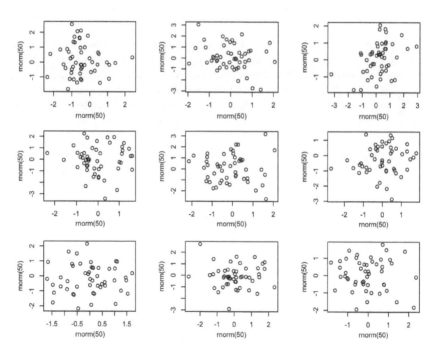

Figure 6.2. Examples of good residual plots (generated via normally distributed random numbers) with 50 data points each

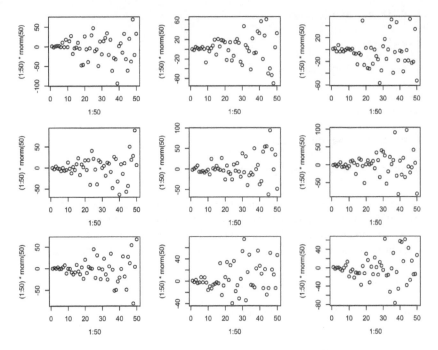

Figure 6.3. Examples of 'bad' residual plots with non-constant variance

It is easy to see patterns in these plots. In saying that these are supposed to be 'good' residual plots, I mean that the visualized data has been generated by a process that is known to be normally distributed with no change in variance (no heteroscedasticity). However, as a result of chance processes, there could always be apparent patterns.

The next code chunk generates bad residual plots with non-constant variance. Again, I recommend executing this command repeatedly to train your eyes.

```
par(mfrow = c(3, 3))

for (i in 1:9) plot(1:50, (1:50) * rnorm(50))
```

The plots shown in Figure 6.3 reveal non-constant variance because, as you move along the fitted values from left to right, the residuals progressively fan out (heteroscedasticity).

6.4. Collinearity

When doing a multiple regression analysis, collinearity describes situations where one predictor can be predicted by other predictors. Collinearity frequently arises from highly correlated predictors, and it makes regression models harder to interpret (see Zuur et al., 2010: 9).

Let's demonstrate collinearity with a familiar example. The following code recreates the randomly generated data from Chapter 4. There is one additional detail: setting a seed value for the random number generation via set.seed() ensures that you and I are working with the same random numbers. In this case, I have chosen the seed value 42 (so as long as you and I use the same number, we will get the same results).

```
set.seed(42)   # seed value for random numbers

x <- rnorm(50)

y <- 10 + 3 * x + rnorm(50)

tidy(lm(y ~ x))
```

```
          term   estimate  std.error  statistic        p.value
1  (Intercept)  10.093852  0.1283994   78.61294   2.218721e-52
2            x   2.807983  0.1125854   24.94091   3.960627e-29
```

The resulting regression model estimates a slope of about +2.81. So far, so good. What if you added a second predictor to the model that was almost exactly the same as the first? To create such a scenario, make a copy of x.

```
x2 <- x
```

Let's change one number of the new x2 vector so that the two vectors are not exactly equal.

```
x2[50] <- -1
```

As expected from two vectors that only differ in one number, there is extreme correlation (Pearson's *r* is 0.98; see Chapter 5).

```
cor(x, x2)
```

```
[1] 0.9793935
```

Given this setup, it comes as no surprise that x2 also predicts y, just as was the case with x.

```
tidy(lm(y ~ x2))
```

```
          term   estimate std.error  statistic        p.value
1  (Intercept)  10.181083 0.1669154   60.99548   3.833676e-47
2           x2   2.724396 0.1457204   18.69605   1.123051e-23
```

What happens if you enter x and x2 together into the same model?

```
xmdl_both <- lm(y ~ x + x2)

tidy(xmdl_both)
```

```
        term   estimate std.error   statistic       p.value
1 (Intercept) 10.0794940 0.1303314 77.3374070 3.352074e-51
2           x  3.2260125 0.5599087  5.7616761 6.164895e-07
3          x2 -0.4255257 0.5582050 -0.7623108 4.496834e-01
```

Notice how much the slope of x2 has changed. It's negative, even though the data has been set up so that x2 and y are positively related. When dealing with strong collinearity, it often happens that the coefficients change drastically depending on which predictors are in the model.

To assess whether you have to worry about collinearity in your analysis, you can use 'variance inflation factors' (VIFs). These measure the degree to which one predictor can be accounted for by the other predictors. There are different recommendations for what constitutes a worrisome VIF value, with some recommending that VIF values larger than 10 indicate collinearity issues (Montgomery & Peck, 1992). Following Zuur et al. (2010), I have used a more stringent cut-off of 3 or 4 in past studies. However, there are also researchers who warn against using variance inflation factors to make decisions about which predictors to include, an issue to which we return below (O'brien, 2007; Morrissey & Ruxton, 2018).

The vif() function from the car package (Fox & Weisberg, 2011) can be used to compute variance inflation factors. For xmdl_both, the variance inflation factors are very high, certainly much in excess of 10, thus indicating strong collinearity.

```
library(car)

vif(xmdl_both)
```

```
       x        x2
24.51677 24.51677
```

Let's compare the variance inflation factors of this model to those of the iconicity model fitted earlier in this chapter.

```
vif(icon_mdl_z)
```

```
   SER_z CorteseImag_z    Syst_z    Freq_z
1.148597      1.143599  1.015054  1.020376
```

For the iconicity model, all variance inflation factors are close to 1, which is good. The actual analysis in presented in Winter et al. (2017) did run into some collinearity issues, specifically caused by the high correlation of concreteness and imageability. We ultimately decided to drop concreteness not only because of high variance inflation factors, but also because it measures a similar theoretical construct as imageability (see Connell & Lynott, 2012). This exclusion was described and justified in the paper. In addition, the supplementary materials showed that our main results were not affected by swapping concreteness in for imageability. Neither were our main conclusions affected by leaving both variables in the same model.

It is important to mention that sample size interacts with collinearity. All else being equal, more data means that regression coefficients can be estimated more precisely (O'brien, 2007; Morrissey & Ruxton, 2018). So, rather than excluding variables from a study, you could also collect more data so that coefficients can be measured precisely even in the presence of collinearity. It is also important to emphasize that collinearity should not be treated as a fault of multiple regression. It simply points to situations where the direct effects are difficult to measure when the predictors are quite entangled.

If you are dealing with a situation where you have rampant collinearity for lots of predictors, regression may not be the answer. There are approaches better suited for finding the most impactful predictors among large sets of potentially collinear variables, such as random forests (Breiman, 2001; for an introduction, see Strobl, Malley, & Tutz, 2009; for a linguistic application, see Tagliamonte & Baayen, 2012). Tomaschek, Hendrix, and Baayen (2018) discuss a variety of alternative approaches to deal with collinearity in linguistic data. However, to truly find out which of multiple underlying factors may cause a result, sometimes it may be necessary to conduct additional experiments (for a linguistic example of this, see Roettger, Winter, Grawunder, Kirby, & Grice, 2014).

It is best to think about collinearity during the planning phase of your study. For example, if there are three highly correlated measures of speech rate (such as 'sentences by second', 'words by second', and 'syllables by second'), you could probably make a theoretically motivated choice about which one is the most appropriate predictor. Including them all into the same model would presumably not advance your theory anyway, and it will make interpreting your model harder.

6.5. Adjusted R^2

Now that we know more about multiple regression, we are in a position to talk about 'adjusted R^2' which you have already seen in various model summaries in previous chapters. Just like R^2, adjusted R^2 measures how much the predictors of a model describe the variance of the response. However, adjusted R^2 is more conservative; it will always be lower than R^2, because it includes a penalizing term that lowers R^2 depending on how many predictors are included in a model.[8] This is done to counteract the fact that adding more predictors to a model always leads to an increased opportunity to capture more variance in the response. Thus, adjusted R^2 is there to counteract an unjust inflation of R^2 due to including too many predictors. This helps to diagnose and prevent 'overfitting', which describes situations when models correspond too closely to the idiosyncratic patterns of particular datasets.

The model summary output generated by the `broom` function `glance()` shows that R^2 and adjusted R^2 correspond very closely to each other in the iconicity model described above. This suggests that there is no problem with overfitting. The presence

---

8 The formula is $R^2_{adj} = 1 - \dfrac{(1 - R^2)(N - 1)}{N - k - 1}$. As you can see, adjusted R^2 is a transformation of R^2. Crucially, the formula contains a term for the number of parameters in a model, k. N represents the number of data points.

of junk predictors would be indicated by an adjusted R^2 value that is much lower than the corresponding R^2 value.

```
glance(icon_mdl_z)
```

```
   r.squared adj.r.squared    sigma  statistic      p.value
1  0.2124559     0.2092545  1.001714 66.36346  9.786184e-50
   df  logLik      AIC      BIC deviance df.residual
1   5 -1402.517 2817.035 2846.415 987.3758         984
```

6.6. Chapter Conclusions

This chapter introduced multiple regression, an extension of simple linear regression. Performing multiple regression allows you to look at the effects of one predictor while holding all other predictors constant. To increase the comparability of regression slopes, it may be useful to standardize predictor variables. We then took a tangent and explored the residuals of the multiple regression model using visual diagnostics. After this, the chapter covered the topic of collinearity, which involves dependencies between the predictors that can make the interpretation of model coefficients difficult. Finally, you have been introduced to adjusted R^2, a more conservative version of R^2 that takes the number of predictor variables into account.

6.7. Exercise

6.7.1. Exercise: Analyzing the ELP Data

Load the 'ELP_full_length_frequency.csv' data into R and fit a regression model where raw response durations (column: RT) are modeled as a function of log_{10} frequency and length (as in Chapter 6.1). Then check the variance inflation factors. Next, check how well your model fits the normality and homoscedasticity assumption.

7   Categorical Predictors

7.1. Introduction

All predictors discussed in this book up to this point were continuous. What if you wanted to know whether a response differed between two or more discrete groups? For example, you may want to show that voice pitch differs by biological sex (female versus male), that vowel duration differs by place of articulation (e.g., /a/ versus /u/), or that reading times depend on grammatical voice (active versus passive). This chapter covers how to model responses as a function of such categorical predictors.

7.2. Modeling the Emotional Valence of Taste and Smell Words

The example is drawn from Winter (2016), where I was interested in the evaluative functions of perceptual adjectives. Although a word's evaluative or 'affective' quality can differ along many important dimensions (Hunston, 2007; Bednarek, 2008), I focused exclusively on the positive/negative dimension, that is, whether a word is overall pleasant or unpleasant. To operationalize this dimension, I used the Warriner et al. (2013) emotional valence ratings that you already explored in Chapter 3.

It has been suggested that smell words are overall more negative (Rouby & Bensafi, 2002: 148–149; Krifka, 2010; Jurafsky, 2014: 96), especially when compared to taste words. This can also be assessed via the contexts in which words occur. For example, the taste word *sweet* collocates with such pleasant nouns as *aroma*, *music*, *smile*, and *dreams*. On the other hand, the smell word *rancid* commonly occurs with such nouns as *smell*, *odor*, *grease*, and *sweat*. The emotional valence scores of these nouns can be used to derive what I call a 'context valence' measure for each adjective (see Winter, 2016; see also Snefjella & Kuperman, 2016). For example, the average noun context valence is more positive for *sweet* (5.7) than for *rancid* (5.1). In this chapter, you will build a linear model to describe the relationship between context valence and sensory modality (taste versus smell).

Figure 7.1 plots the noun context valence measure for all the taste words (diamond shapes) and smell words (circles).[1] This figure shows that taste words tend to occur in more positive contexts than smell words, although there is considerable overlap between the two distributions.

1 The sensory modality classifications are taken from Lynott and Connell (2009).

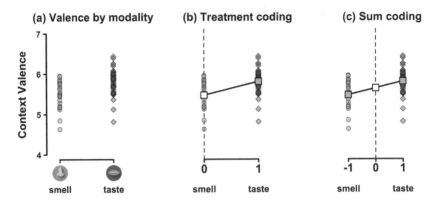

Figure 7.1. (a) Noun context valence for taste and smell words; (b) treatment coding, with smell and taste words positioned at 0 and 1, respectively; (c) sum-coding, with smell = −1 and taste = +1; white squares represent intercepts

Regression works well with sets of numbers. To incorporate the categorical factor 'modality' into a regression model, the labels 'taste' and 'smell' are replaced with numerical identifiers, a process which is called 'dummy coding'. Geometrically, assigning numbers to categories means that the categories are placed into a coordinate system, which is depicted in Figure 7.1b. In this case, smell words are located at $x = 0$, and taste words at $x = 1$.

The particular coding system used in Figure 7.1b is called 'treatment coding'.[2] The overarching term used for treatment coding and other coding systems is 'contrasts'. Within the treatment coding system, the category at $x = 0$ is called the reference level, and it will assume the role of the intercept of the regression model. The intercept is represented by the white square in Figure 7.1b, which is the mean of the smell words in this case.

Now, remember two facts from the previous chapters: first, the mean is the value that is closest to all data points for a univariate dataset. Second, regression attempts to minimize the residuals. Taken together, these two facts entail that when a regression model is fitted onto a variable with two categories, the line has to go through the means of both categories. The regression line shown in Figure 7.1b can only be interpreted at the discrete points $x = 0$ and $x = 1$. Predictions generated for intermediate values do not make any sense.

Remember that a slope can be paraphrased as 'rise over run' (Chapter 4). So, for Figure 7.1b, what's the 'rise'? As you move from $x = 0$ to $x = 1$, you ascend by +0.3 emotional valence points, the difference between the two means. What's the corresponding 'run'? As you only move from $x = 0$ to $x = 1$, the run is exactly one unit. Applying the 'rise over run' formula then yields $\frac{0.3}{1} = 0.3$. Thus, the slope of the modality predictor is *exactly* equal to the mean difference between the two

2 'Dummy coding' is often used synonymously with 'treatment coding'.

groups. This means that, for categorical predictors, your regression slopes are actually differences between groups! The predictive equation for the data shown in Figure 7.1b is the following:

$$\underset{\text{(smell words)}}{context\ valence} = \quad 5.5 \quad + \quad \underset{\text{(change from smell to taste)}}{0.3 * modality} \tag{E7.1}$$

Let's plug in the corresponding dummy codes. When *modality* = 1, you get $5.5 + 0.3*1 = 5.8$. When *modality* = 0, you get $5.5 + 0.3*0 = 5.5$. You can think of the latter as *not* applying the change from taste to smell.

The choice of the reference level is up to the user. There's nothing that stops you from making 'taste' the reference level rather than 'smell'. In that case, your equation would look like this:

$$context\ valence = 5.8 + (-0.3) * modality \tag{E7.2}$$

You can think of E7.1 and E7.2 as two perspectives on the same data. You either view the data from the perspective of smell words, looking 'up' towards taste words (E7.1), or you view the data from the perspective of taste words, looking 'down' towards smell words (E7.2). The question of assigning reference levels is only a question of representation.[3]

7.3. Processing the Taste and Smell Data

Let's begin by loading in the relevant data.

```
library(tidyverse)
library(broom)

senses <- read_csv('winter_2016_senses_valence.csv')

senses
```

```
# A tibble: 405 x 3
    Word       Modality    Val
    <chr>      <chr>       <dbl>
1 abrasive   Touch       5.40
2 absorbent  Sight       5.88
3 aching     Touch       5.23
4 acidic     Taste       5.54
5 acrid      Smell       5.17
6 adhesive   Touch       5.24
7 alcoholic  Taste       5.56
```

---

3 Harkening back to the discussion in Chapter 5, changing the reference level is a linear transformation, as it leaves the relationship between the data points untouched.

```
 8 alive      Sight     6.04
 9 amber      Sight     5.72
10 angular    Sight     5.48
# ... with 395 more rows
```

This data pairs the Lynott and Connell (2009) modality classifications with the context valence measure from Winter (2016).[4] It's a good idea to spend some time familiarizing yourself with this data. For example, you could tabulate the content of the `Modality` column to see how many words there are per sensory modality (see Chapter 2). It's also a good idea to explore the `range()`, `mean()`, and `sd()` of the `Val` column. In addition, histograms may be useful to get an overview. I will skip these steps now for the sake of discussing how to deal with categorical predictors, but you are welcome to have a thorough look at this dataset before you continue.

Let's reduce the tibble to a subset containing only taste and smell words.

```
chem <- filter(senses, Modality %in% c('Taste', 'Smell'))
```

It's a good idea to verify that this has actually worked. The following command tabulates the number of words per sensory modality. As you can see, there are only taste and smell words.

```
table(chem$Modality)
```

```
Smell Taste
   25    47
```

Let's compute the means and standard deviations for each category. For this, use `group_by()` together with `summarize()`. Grouping the tibble ensures that the `summarize()` function knows what groups to summarize by.

```
chem %>% group_by(Modality) %>%
  summarize(M = mean(Val), SD = sd(Val))
```

```
# A tibble: 2 x 3
  Modality      M     SD
  <chr>     <dbl>  <dbl>
1 Smell      5.47  0.336
2 Taste      5.81  0.303
```

These values would be good to report in a paper. For example, you might want to write a statement such as the following: 'The average context valence of taste words was higher ($M = 5.81$, $SD = 0.30$) than the average context valence of smell words ($M = 5.47$, $SD = 0.34$).'

---

4 To construct the context valence measure, I used the emotional valence ratings by Warriner et al. (2013) and the Corpus of Contemporary American English (Davies, 2008).

How could you visualize this difference? One option is a box-and-whiskers plot (see Chapter 3.6). For the following `ggplot2`, the categorical predictor is mapped to the *x*-values, and the continuous response is mapped to the *y*-values. You can additionally map the categorical predictor to the `fill` argument, which fills the boxes according to `Modality`. The additional command `scale_fill_brewer()` adds a "Color Brewer" palette. There are numerous palettes to explore on http://colorbrewer2.org. The palette picked here (called 'PuOr', which stands for 'PurpleOrange') is 'photocopy safe', so that the two colors appear recognizably different from each other when printed in black and white. The resulting boxplot is shown in Figure 7.2 (left plot).

```
chem %>% ggplot(aes(x = Modality, y = Val, fill = Modality)) +
  geom_boxplot() + theme_minimal() +
  scale_fill_brewer(palette = 'PuOr')
```

Alternatively, you could create a density graph (see Figure 7.2, right plot). This is essentially a smoothed version of a histogram (using what's called 'kernel density estimation', which won't be explained here). The `alpha` of the density graph is set to 0.5 to make the filling of the density curves transparent (play with different `alpha` values to see how this changes the look and feel of this plot).

```
chem %>% ggplot(aes(x = Val, fill = Modality)) +
  geom_density(alpha = 0.5) +
  scale_fill_brewer(palette = 'PuOr')
```

Now that you have a thorough understanding of this dataset, you are in a good position to perform a linear regression analysis.

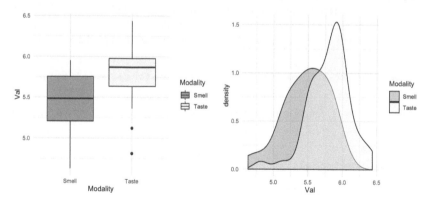

Figure 7.2. Left: boxplot of the emotional valence difference between taste and smell words; right: kernel density plot of the same data

7.4. Treatment Coding in R

Let us fit a regression model where valence is modeled as a function of modality.

```
chem_mdl <- lm(Val ~ Modality, data = chem)

tidy(chem_mdl) %>% select(term, estimate)
```

```
          term  estimate
1   (Intercept) 5.4710116
2  ModalityTaste 0.3371123
```

Notice that the estimate for the modality predictor is shown as `ModalityTaste`, even though you used just 'Modality' in the model formula. This always happens with categorical predictors that are treatment-coded. Whatever category is mentioned in the slope of the output (here: `Taste`) corresponds to the '1' in the treatment coding scheme. This means that the category that is *not* explicitly mentioned in the coefficient output corresponds to the '0' and is hidden in the intercept. In this case, the intercept is the average context valence of smell words. The `ModalityTaste` slope then represents a change from the intercept (`Smell`) *towards* `Taste`.

Checking the fitted values shows that this model only makes two predictions, one for each category.

```
head(fitted(chem_mdl))
```

```
        1         2         3         4         5         6
5.808124  5.471012  5.808124  5.471012  5.471012  5.808124
```

Since the model only makes two predictions (that are repeated for each data point, depending on whether it is a taste or a smell word), you can easily 'read off' what this model predicts from the output of the `fitted()` function. Nevertheless, because this will be more useful when dealing with complex models, let's use `predict()` to generate the predictions instead. You have first encountered the `predict()` function in Chapter 4. Start by creating a data frame or tibble to generate predictions for.

```
chem_preds <- tibble(Modality = unique(chem$Modality))
```

The `unique()` function is used here to reduce the `Modality` column to the unique types.[5]

```
unique(chem$Modality)
```

```
[1] "Taste" "Smell"
```

---

5 The `unique()` function is quite similar to `levels()`. However, the latter can only be used for factor vectors. Currently, the `Modality` column is a character vector.

Finally, let's compute the fitted values and append them to the chem_preds tibble.

```
chem_preds$fit <- predict(chem_mdl, chem_preds)

chem_preds
```

```
# A tibble: 2 x 2
  Modality      fit
  <chr>       <dbl>
1 Taste        5.81
2 Smell        5.47
```

These predictions correspond to what was discussed earlier in the chapter.

7.5. Doing Dummy Coding 'By Hand'

You would usually not do this in an actual data analysis, but for pedagogical reasons it helps to create the dummy codes yourself. The following code creates a new column Mod01 using ifelse(). This function spits out '1' if the statement 'Modality == 'Taste'' is TRUE, and '0' if it is FALSE.

```
chem <- mutate(chem,
        Mod01 = ifelse(Modality == 'Taste', 1, 0))
```

Let's check that this produced the desired result:

```
select(chem, Modality, Mod01)
```

```
# A tibble: 72 x 2
Modality  Mod01
   <chr>  <dbl>
 1 Taste      1
 2 Smell      0
 3 Taste      1
 4 Smell      0
 5 Smell      0
 6 Taste      1
 7 Taste      1
 8 Taste      1
 9 Taste      1
10 Taste      1
# ... with 62 more rows
```

You can now fit a linear model with the new Mod01 column as predictor.

```
lm(Val ~ Mod01, data = chem)
```

```
Call:
lm(formula = Val ~ Mod01, data = chem)

Coefficients:
(Intercept)      Mod01
     5.4710     0.3371
```

Notice that the resulting numbers are *exactly* the same as from the model discussed earlier. This highlights how a regression with a categorical predictor is the same as regressing a continuous variable on a set of 0s and 1s.

7.6. Changing the Reference Level

What if you wanted to change the reference level? For this, it helps to convert the Modality column into factor vector. As discussed in Chapter 2.2, the beauty of read_csv() and tibbles is that text data is stored in terms of character vectors, which are much easier to manipulate than factor vectors. However, when you want to have control of contrast coding schemes, you may have to work with factor vectors.

The following mutate() command takes the Modality column and converts it into a factor vector, which is then immediately releveled so that 'Taste' becomes the reference level.[6] The releveled factor is stored in the column ModRe. Notice that this second step only works because the column has at this second stage in the mutate() command already been changed to a factor.[7]

```
chem <- mutate(chem,
         Modality = factor(Modality),
         ModRe = relevel(Modality, ref = 'Taste'))
```

Let's check the levels of both factors with levels().

```
levels(chem$Modality)
```

```
[1] "Smell" "Taste"
```

```
levels(chem$ModRe)   # releveled factor
```

```
[1] "Taste" "Smell"
```

Whatever is mentioned first in this output is the reference level. Thus, the reference level of the Modality column is 'Smell'; the reference level of the releveled ModRe column is 'Taste'. If nothing else is specified, R will sort the levels

---

6 The factor() function works just like as.factor() in this case.

7 Rather than using relevel(), you can also define the order of levels when creating the factor. Whatever is mentioned first in the levels argument below is made the reference level.

```
chem <- mutate(chem,
          ModRe = factor(Modality, levels = c('Taste','Smell')))
```

alphanumerically, which is why 'Smell' is the reference level of the Modality column.

Let's fit the model with the releveled ModRe predictor.

```
lm(Val ~ ModRe, data = chem) %>%
  tidy %>%
  select(term, estimate)
```

```
          term     estimate
1 (Intercept)    5.8081239
2  ModReSmell   -0.3371123
```

You can see that the intercept is now set at the valence mean of the taste words (5.8). The slope has reversed sign, because it is now the change from taste to smell, which is –0.3.

There's no mathematical reason to use either taste or smell words as reference level. The choice is up to the researcher! So, choose whatever is most intuitive to you in the context of your analysis.

7.7. Sum-coding in R

Besides treatment coding, 'sum-coding' is another commonly used coding scheme. Why would you want to use one coding system over another? You will see in Chapter 8 that, once there are interactions in a model, sum-coding may confer some interpretational advantages. For now, you are going to do a 'dry run', rehearsing this coding system in a situation where both treatment coding and sum-coding are equally good. As you progress through the book, you will see that it is important to have flexibility about representing one's categorical predictors.

When converting a categorical predictor into sum-codes, one category is assigned the value –1; the other is assigned +1, as visualized in Figure 7.1c above. With this coding scheme, the intercept is in the middle of the two categories, which is the conceptual analog of 'centering' for categorical predictors. The y-value of the intercept is now the mean of the means. In other words, the intercept is halfway in between the two categories.[8]

Let's apply the 'rise over run' formula. The 'rise' along the y-axis is still the same, since the mean difference between taste and smell words hasn't changed. However, the 'run' along the x-axis has changed. Jumping from one category (–1) to another (+1) results in an overall change of 2. This means that the slope becomes $\frac{0.3}{2} = 0.15$, which is *half* the difference of the means. You can think of this as sitting at 0 in the middle of the two categories. In one direction, you look 'up' to the taste words. In the other direction, you look 'down' to the smell words.

Let's sum-code the Modality column. Just in case, the following code repeats the conversion to factor.

8 If the data is balanced (there is an exactly equal number of data points in each category), the intercept will also be the overall mean of the dataset.

```
chem <- mutate(chem, Modality = factor(Modality))

# Check:

class(chem$Modality) == 'factor'
```

```
[1] TRUE
```

Let's check what coding scheme R assigns to the Modality factor by default. This can be interrogated with the contrasts() function.

```
contrasts(chem$Modality)
```

```
        Taste
Smell      0
Taste      1
```

When a factor with this coding scheme is internally sent to the linear model function, a new variable Taste is created, which assumes 0 for smell words and 1 for taste words. This is why the output in the regression model was displayed as ModalityTaste, rather than just Modality.

As mentioned above, R uses the treatment coding scheme (0/1) by default. The contr.treatment() function can be used to create this coding scheme explicitly. This function has one obligatory argument: the number of categories that you want the treatment coding scheme for. Let's see how the coding scheme for a binary category looks like.

```
contr.treatment(2)
```

```
   2
1  0
2  1
```

The matrix that is generated by this function tells you about the assignment of categories to numerical identifiers. Specifically, the first category (the row with a 1) is mapped to 0; the second category (the row with a 2) is mapped to 1. The column is named '2', as the dummy variable will be named after the second category (as was the case with ModalityTaste below).

For sum-coding, use contr.sum(). As you can see, when this function is run for two levels, the first category is mapped to 1; the second category is mapped to –1. This time around, the column doesn't have a special name.

```
contr.sum(2)
```

```
   [,1]
1   1
2  -1
```

Let's fit a regression model with a sum-coded Modality factor. First, create a copy of the Modality column. Then recode the coding scheme to sum-codes.

```
chem <- mutate(chem, ModSum = Modality)

contrasts(chem$ModSum) <- contr.sum(2)

lm(Val ~ ModSum, data = chem) %>%
  tidy %>% select(term, estimate)
```

```
          term     estimate
1 (Intercept)    5.6395677
2      ModSum1   -0.1685562
```

Notice that, compared to the previous model, the slope has halved. Moreover, the intercept is now equal to the mean of the means, which can be verified as follows:

```
chem %>% group_by(Modality) %>%
  summarize(MeanVal = mean(Val)) %>%
  summarize(MeanOfMeans = mean(MeanVal))
```

```
# A tibble: 1 x 1
  MeanOfMeans
        <dbl>
1        5.64
```

Notice that the modality slope is represented as 'ModSum1'. Using the '1' after the predictor name is a notational convention for representing the slopes of sum-coded predictors in R.

The following equation uses the coefficients from the sum-coded model to derive predictions for both categories.

$$
\begin{aligned}
taste\,valence &= 5.6 + (-0.2)*(+1) = 5.4 \\
smell\,valence &= 5.6 + (-0.2)*(-1) = 5.8
\end{aligned}
\tag{E7.3}
$$

Notice that, barring some differences due to rounding, the predictions are exactly the same as those of the treatment-coded model.

7.8. Categorical Predictors with More Than Two Levels

The example discussed so far only pertains to a binary categorical predictor. But what if your predictor variable has more than two levels? Let's go back to the senses tibble, which contains words for all of the five senses (sight, touch, sound, taste, and smell).

```
unique(senses$Modality)
```

```
[1] "Touch" "Sight" "Taste" "Smell" "Sound"
```

Fit a linear model with this five-level predictor to see what happens.

```
sense_all <- lm(Val ~ Modality, data = senses)

tidy(sense_all) %>% select(term:estimate) %>%
  mutate(estimate = round(estimate, 2))
```

```
          term  estimate
1   (Intercept)     5.58
2 ModalitySmell    -0.11
3 ModalitySound    -0.17
4 ModalityTaste     0.23
5 ModalityTouch    -0.05
```

The regression output shows four slopes, namely, four differences to one shared reference level.[9] Newcomers to regression modeling are often disappointed about the fact that a model's coefficient table only presents a very partial view of the differences between the categories in a study. Often, researchers are interested in some form of comparison between all categories. There are three answers to this concern. First, observe the fact that, with one reference level and four differences, you can actually compute predictions for all five categories, as will be demonstrated below. Second, it is possible to test the overall effect of the five-level predictor, as will be demonstrated in Chapter 11. Third, it is possible to perform tests for all pairwise comparisons, which will also be demonstrated in Chapter 11. For now, I want you to accept the fact that your regression model represents five categories in terms of a reference level and 'only' four differences.

In the case of the present dataset, what is the reference level? You can determine the reference level from the output: it is whatever category is *not* shown as one of the slopes. Since you haven't specified the reference level yourself, R takes whatever comes first in the alphabet, which happens to be Sight in this case. Thus, the sight words are 'hidden' in the intercept. The first slope, ModalitySmell, then is the difference between sight and smell words. The sign of this coefficient is negative, so smell words are more negative than sight words.

To generate the model equation from the R output, read off the estimate column from top to bottom and add all terms together:

$$Valence = 5.58 + (-0.11) * Smell + (-0.17) * Sound$$
$$+ 0.23 * Taste + (-0.05) * Touch \tag{E7.4}$$

To get the prediction for, say, the smell words, plug in 1 for the 'Smell' variable, and 0 for all others: 5.58 + (−0.11)*Smell = 5.47. Next, 5.58 + (−0.17)*Sound is the predicted context valence for sound words, and so on. To retrieve the average valence

---

9 This provides a view of the corresponding treatment coding scheme:

```
contr.treatment(5)
  2 3 4 5
1 0 0 0 0
2 1 0 0 0
3 0 1 0 0
4 0 0 1 0
5 0 0 0 1
```

of sight words, you simply need to read off the intercept. This is the only mean that can be read off from the coefficient table without any arithmetic.

To save yourself performing these computations per hand, you always have the predict() function at your disposal.

```
sense_preds <- tibble(Modality =
                sort(unique(senses$Modality)))

sense_preds$fit <- round(predict(sense_all, sense_preds), 2)

sense_preds
```

```
# A tibble: 5 x 2
  Modality    fit
  <chr>     <dbl>
1 Sight      5.58
2 Smell      5.47
3 Sound      5.41
4 Taste      5.81
5 Touch      5.53
```

In this sequence of codes, sort() is wrapped around the output of unique() to put the different modalities into alphabetical order. The output of this is saved into the fit column of the sense_preds tibble. The result shows that the predict() function derived the prediction for each of the five levels. These values *aren't* relative to the sight reference level: these are the actual predictions for each category.

7.9. Assumptions Again

Let's use the sense_all model to learn something more about residuals and the regression assumptions that relate to the residuals. The code below reproduces the visual diagnostics plots discussed in Chapter 6. The resulting plots should look similar to Figure 7.3.

```
par(mfrow = c(1, 3))

# Plot 1, histogram:

hist(residuals(sense_all), col = 'skyblue2')

# Plot 2, Q-Q plot:

qqnorm(residuals(sense_all))
qqline(residuals(sense_all))

# Plot 3, residual plot:

plot(fitted(sense_all), residuals(sense_all))
```

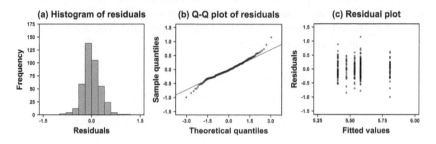

Figure 7.3. (a) Histogram; (b) Q-Q plot; and (c) residual plot to assess the normality and constant variance assumptions for the `sense_all` model

The model seems to conform fairly well to the normality assumption: the distribution of residuals looks very normal in the histogram, and the Q-Q plot also indicates a good fit with the normal distribution (you will often find that the residuals fan out a tiny bit from the Q-Q line for more extreme values). However, the residual plot might look weird to you. Are these residuals consistent with the constant variance assumption? And where do these vertical stripes come from? The stripes are there because the `sense_all` predicts five fitted values, one for each category. This is why there are no in-between values. What matters for the assessment of homoscedasticity ('equal variance') is that the residuals in Figure 7.3c are about equally spread out for each of the categories, which seems to be the case for this dataset.

7.10. Other Coding Schemes

There are many more coding schemes not discussed in this book. For example, the 'Helmert coding' and 'forward difference coding' schemes may become useful when you want to incorporate ordered categorical predictors into your model. Helmert coding compares the levels of a variable with the mean of the subsequent levels of that variable. This coding scheme is implemented in R via the `contr.helmert()` function. For a factor with four levels, this coding system looks like this.

```
contr.helmert(4)

  [,1] [,2] [,3]
1   -1   -1   -1
2    1   -1   -1
3    0    2   -1
4    0    0    3
```

In the corresponding regression output, the first slope indicates the difference between levels 1 and 2. The second slope indicates the difference between levels 1 and 2, compared to level 3. The third slope indicates the difference between levels 1, 2, and 3, compared to level 4. Thus, each consecutive level is compared to the mean of all previous levels in an ordered sequence. This is useful, for example, when testing ordered predictors such as education level (PhD > MA > BA, etc.).

The two coding schemes that form the focus of this chapter, treatment coding and sum-coding, will suit most of your needs. When you report the results of a regression model with categorical predictors, it's important to mention what coding scheme you used so that the reader can interpret any coefficient estimates you mention in the paper. It's good to be explicit about this, and I recommend doing this even when you use R's default coding system (treatment coding).

7.11. Chapter Conclusions

This chapter covered how to incorporate categorical predictors into a regression framework. This allows fitting models with both continuous and categorical predictors. The only novelty compared to the continuous case is that you have to be careful in interpreting the output of categorical models, keeping track of how categories are converted into dummy codes. This chapter covered two coding schemes: treatment coding and sum-coding.

7.12. Exercises

7.12.1. Plotting Categorical Differences

Using the `senses` tibble, create a boxplot that shows the valence for all the five senses. Next, plot a density graph with `geom_density()`. Map the `Modality` column onto the fill argument and increase the alpha to get transparent colors. What happens if you add the following new layer to the plot?

```
+ facet_wrap(~Modality)
```

7.12.2. Iconicity as a Function of Sensory Modality

Let's assess the degree to which perceptual words differ in terms of iconicity as a function of sensory modality, as explored in Winter et al. (2017) (see Chapter 2). The following code loads in the Lynott and Connell (2009) modality ratings for adjectives and our iconicity ratings. The two tibbles are then merged, and a subset of the columns is extracted using `select()`.

```
lyn <- read_csv('lynott_connell_2009_modality.csv')
icon <- read_csv('perry_winter_2017_iconicity.csv')

both <- left_join(lyn, icon)

both <- select(both, Word, DominantModality, Iconicity)

both
```

```
# A tibble: 423 x 3
   Word       DominantModality Iconicity
   <chr>      <chr>                 <dbl>
 1 abrasive   Haptic                 1.31
 2 absorbent  Visual                0.923
 3 aching     Haptic                 0.25
 4 acidic     Gustatory             1
 5 acrid      Olfactory             0.615
 6 adhesive   Haptic                 1.33
 7 alcoholic  Gustatory             0.417
 8 alive      Visual                 1.38
 9 amber      Visual                 0
10 angular    Visual                 1.71
# ... with 413 more rows
```

Fit a linear model where `Iconicity` is modeled as a function of the categorical predictor `DominantModality`. Interpret the output. What does the intercept represent? Can you use the coefficients to derive predictions for all five categories? Compare your results against the descriptive means, for which you can use `group_by()` and `summarize()`.

8 Interactions and Nonlinear Effects

8.1. Introduction

An interaction describes a situation where the influence of a predictor on the response depends on another predictor. That is, interactions are about relationships between predictors, and how two or more predictors *together* influence a response variable. McElreath (2016: 120) provides an intuitive example of an interaction: imagine modeling plant growth as a function of the two predictors 'water use' and 'sun exposure'. Neither of these predictors alone will have a great impact on plant growth. Only if there is *both* water and sun will the plant grow. That is, the influence of the sun exposure predictor critically depends on the water use predictor, and vice versa.

In the language sciences, interactions are ubiquitous. For example, in Winter and Bergen (2012), we asked English speakers to read sentences that differed in descriptions of distance, such as *You are looking at the beer bottle right in front of you* (near) versus *You are looking at the beer bottle in the distance* (far). After having read one of these two sentences, participants either saw a large image of a bottle (near) or a small image (far). We measured the speed with which participants verified whether the picture object was mentioned in the sentence. In our model, the two predictors were 'sentence distance' and 'picture size'. Crucially, these predictors interacted with each other: looking at a large image after reading a 'near' sentence was faster, and so was looking at a small image after reading a 'far' sentence. In contrast, mismatching pairs of predictor levels (near sentence/small image; far sentence/large image) resulted in slower response times. This is an example of an interaction, because it is specific combinations of the predictor levels that matter.

The following shows the equation of a model with two predictor variables, x_1 and x_2 (without error term):

$$y = b_0 + b_1 x_1 + b_2 x_2 \qquad (E8.1)$$

To incorporate an interaction, the two predictors are multiplied by each other $(x_1 * x_2)$. Regression will then estimate the corresponding slope for this new predictor, b_3. The numerical value of this slope describes the strength of this 'multiplicative effect'. When b_3 is estimated to be close to 0, the interaction is weak. The further way b_3 is from 0, the stronger the interaction effect.

$$y = b_0 + b_1 x_1 + b_2 x_2 + b_3 (x_1 * x_2) \qquad (E8.2)$$

In psychology, interactions are sometimes called 'moderator variables' because they 'moderate' the effects of other predictors. You can think of it this way: by multiplying the two predictors with each other, you effectively 'interlock' them, and the coefficient b_3 specifies *how* the two predictors are interlocked.

This will become clearer with examples. This chapter introduces you to different types of interactions. First, interactions between a categorical and a continuous variable. Then interactions between two categorical variables. Finally, interactions between two continuous variables. The final section of this chapter considers another 'multiplicative' type of model: polynomial regression.

However, before you embark on the journey that is this chapter, I want to prepare for what's coming up. Interpreting interactions is very hard and takes time. There's also going to be a bit more arithmetic than in previous chapters. If you are new to regression modeling, this chapter may be worth reading more than once.

8.2. Categorical * Continuous Interactions

Let us revisit the iconicity model from Chapter 6. To remind you: iconicity describes the degree to which a word form resembles the meaning of a word. For example, onomatopoeic words such as *bang* and *beep* are iconic because they imitate the sounds these words describe. One outcome of the analysis presented in Chapter 6 was that words with more sensory content were on average more iconic than words with less sensory content (see Winter et al., 2017).

Figure 8.1 shows the relationship between iconicity and sensory experience ratings (SER) separately for nouns and verbs. The lines inside the plots show the linear model fits of simple bivariate regression models (`Iconicity ~ SER`), one for nouns and one for verbs. You can easily see that the relationship between sensory experience and iconicity is stronger for verbs than for nouns. The respective linear regression models estimate the SER slopes to be +0.63 for verbs and +0.12 for nouns. The fact that SER has a different effect on iconicity depending on part of speech hints at an interaction, but you cannot simply compare the slopes of separate models (for a nice discussion of this, see Vasishth & Nicenboim, 2016). Instead, the interaction should be modeled explicitly.

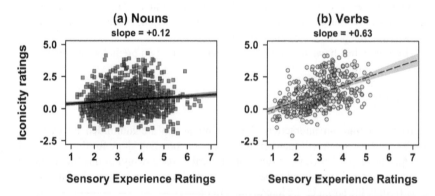

Figure 8.1. The relationship between sensory experience ratings and iconicity for (a) nouns and (b) verbs; notice the steeper slope for verbs; the shaded 95% confidence regions will be explained in Chapters 9–11

First, begin by loading the data from Chapter 5 into R.

```
library(tidyverse)
library(broom)

icon <- read_csv('perry_winter_2017_iconicity.csv')

icon
```

```
# A tibble: 3,001 x 8
    Word       POS         SER   CorteseImag Conc  Syst  Freq
    <chr>      <chr>       <dbl>       <dbl> <dbl> <dbl> <int>
 1  a          Grammati…   NA             NA  1.46    NA 1.04e6
 2  abide      Verb        NA             NA  1.68    NA 1.38e2
 3  able       Adjective   1.73           NA  2.38    NA 8.15e3
 4  about      Grammati…   1.2            NA  1.77    NA 1.85e5
 5  above      Grammati…   2.91           NA  3.33    NA 2.49e3
 6  abrasive   Adjective   NA             NA  3.03    NA 2.30e1
 7  absorbe…   Adjective   NA             NA  3.1     NA 8.00e0
 8  academy    Noun        NA             NA  4.29    NA 6.33e2
 9  accident   Noun        NA             NA  3.26    NA 4.15e3
10  accordi…   Noun        NA             NA  4.86    NA 6.70e1
# … with 2,991 more rows, and 1 more variable:
#   Iconicity <dbl>
```

Most columns are irrelevant for the current analysis. All you need to proceed is the columns for Iconicity (response), SER (predictor 1), and POS (predictor 2).

Let's use unique() to have a look at the names of the different lexical categories ('parts of speech').

```
unique(icon$POS)
```

```
[1] "Grammatical"  "Verb"       "Adjective"  "Noun"
[5] "Interjection" "Name"       "Adverb"     NA
```

Use table() to check how many words there are per parts of speech. If you wrap sort() around the table, the categories will be sorted in terms of ascending counts. Notice that there are many more nouns than verbs.

```
sort(table(icon$POS))
```

```
     Name Interjection    Adverb Grammatical
       15           17        39          80
Adjective         Verb      Noun
      535          557      1704
```

Let's filter the tibble so that it includes only nouns and verbs.

```
NV <- filter(icon, POS %in% c('Noun', 'Verb'))
```

Check that this has produced the desired effect:

```
table(NV$POS)
```

```
Noun Verb
1704  557
```

The `table()` command shows that there are now only nouns and verbs in this dataset.

For pedagogical reasons, the first model considered here regresses iconicity on SER and part of speech *without* including an interaction term.

```
NV_mdl <- lm(Iconicity ~ SER + POS, data = NV)

tidy(NV_mdl) %>% select(term, estimate)
```

```
          term     estimate
1 (Intercept)   -0.1193515
2         SER    0.2331949
3     POSVerb    0.6015939
```

As always, you should spend considerable time interpreting the coefficients. First, the intercept is the prediction for nouns with 0 sensory experience ratings. You know that nouns are in the intercept (reference level) because it says 'POSVerb' in the output, and because 'n' comes before 'v' in the alphabet. The positive coefficient (+0.60) thus shows that verbs are more iconic than nouns.

Next, the sensory experience slope (SER) is estimated to be +0.23, which indicates a positive relationship between iconicity and sensory experience. Remember that the SER slope of the noun model discussed was +0.12; for verbs, the slope was +0.63. The slope +0.23 of this model is in between the slopes for nouns and verbs, and it is closer to the nouns because there are more nouns than verbs. This model does 'know' that the SER slope could be different for nouns and verbs, and in ignoring this interaction the model ends up mischaracterizing the SER slope for both lexical categories.

Figure 8.2 visualizes the NV_mdl model. The white square is the intercept at SER = 0. The intercept is at the noun category, and verbs (dashed line) have higher iconicity than nouns. As the lines are parallel, this is regardless of which value of SER you consider.

Figure 8.3a shows a more complex model which includes an interaction term. The most striking difference to Figure 8.2 is the fact that the two lines are not parallel anymore. This means that sensory experience is estimated to have different effects on iconicity for nouns and verbs. Likewise, it means that the degree to which nouns and verb differ from each other in terms of iconicity also depends on what SER value one considers. Thus, you cannot interpret each of the predictors in isolation anymore. The two predictors are conditioned on each other and have to be interpreted together.

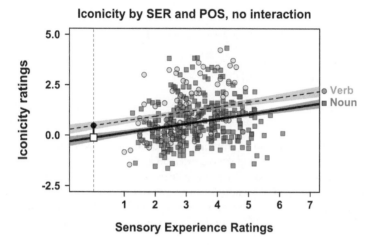

Figure 8.2. Iconicity as a function of sensory experience ratings; the regression lines represent a model that does *not* include an interaction term 'Iconicity ~ SER + POS'; the white square represents the intercept of this model; to increase clarity, only a random 25% of the data points are shown for each group

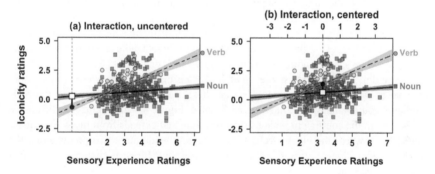

Figure 8.3. (a) The plot corresponding to the model 'Iconicity ~ SER * POS'; (b) the plot corresponding to the model 'Iconicity ~ SER_c * POS' with a centered sensory experience predictor; notice the shift in the intercept (white square), which means that the verb-noun difference is evaluated elsewhere; graph inspired by Schielzeth (2010)

To fit the model shown in Figure 8.3, you need to include an interaction term in your model formula, which is done via the asterisk '*'.[1]

---

1 There are two alternative ways of specifying interactions in model formulas:
```
lm(iconicity ~ SER * POS, data = NV)
# Same as:
lm(iconicity ~ SER + POS + SER:POS, data = NV)
```
The latter effectively 'spells out' the more compressed 'SER * POS' notation and highlights that the interaction involves a third term in the model, SER:POS.

```
NV_int_mdl <- lm(Iconicity ~ SER * POS, data = NV)

tidy(NV_int_mdl) %>% select(term, estimate)
```

```
          term     estimate
1 (Intercept)    0.2739423
2         SER    0.1181651
3     POSVerb   -0.9554158
4 SER:POSVerb    0.5083802
```

As always, devote time to interpreting these coefficients. When there are interactions in your model, you need to devote *extra* time.

First, let's look at how many coefficients there are altogether: there are two coefficients for each predictor (SER and POSVerb), as well as a separate coefficient for the interaction (SER:POSVerb). Second, remind yourself about what's in the intercept. As was the case in the model before, the intercept is the prediction for nouns with 0 SER.

Crucially, now that the lines are not parallel anymore, the meaning of the SER and POSVerb coefficients have changed. The SER slope is now the slope of sensory experience *only for the nouns*. A common thinking trap is to think that the SER coefficient is the average SER slope, which is *not* the case. In fact, notice that the slope for SER is the same as was the case for the noun-only model discussed above (+0.12).

Likewise, the POSVerb effect is the noun-verb difference *only for words with 0 sensory experience*. Compare Figure 8.3a to the coefficient table and notice how, at the intercept (white square), nouns are actually *more* iconic than verbs (POSVerb = –0.96). However, don't be misled to suggest that this means that nouns are overall more iconic than verbs—if at all, the opposite is true. It all depends on *where* along the SER predictor you evaluate the noun-verb difference. Figure 8.3 highlights the difference that's seen in the coefficient table as the black line that extends from the white square (intercept) towards the black circle. If you mentally slide these two points along the two regression lines, you notice how the noun-verb difference changes and in fact even reverses sign.

How can you interpret the coefficient of the interaction term SER:POSVerb? In this context, it is appropriate to think of this coefficient as a 'slope adjustment term'. The coefficient is +0.51, which means that the SER slope is steeper for verbs. Adding the +0.51 interaction term to the slope of the nouns +0.12 yields the slope of the verbs, which is +0.63. Put differently, the verbs get an additional boost with respect to the SER effect.

The interpretation of models with interactions is often greatly facilitated when continuous variables are centered. Remember from Chapter 5 that centering a continuous predictor sets the intercept to the mean. If you center sensory experience ratings, you put the intercept to the center of mass of the data points, as shown in Figure 8.3b. Notice how for this central SER value, verbs are actually *more* iconic than nouns. This is arguably a more adequate characterization of the noun-verb difference than evaluating this difference at a sensory experience rating of 0, which doesn't even exist in the data because the scale started at 1. Let's recompute the model with a centered SER variable.

```
# Center SER:

NV <- mutate(NV, SER_c = SER - mean(SER, na.rm = TRUE))

# Fit model with centered predictor:

NV_int_mdl_c <- lm(Iconicity ~ SER_c * POS, data = NV)

# Check coefficients:

tidy(NV_int_mdl_c) %>% select(term, estimate)
```

```
          term   estimate
1   (Intercept)  0.6642298
2        SER_c  0.1181651
3      POSVerb  0.7237133
4 SER_c:POSVerb  0.5083802
```

Centering sensory experience ratings has reversed the sign of the POSVerb coefficient, which is now +0.72. This now has a more meaningful interpretation: it is the difference between nouns and verbs for words with average sensory experience ratings, rather than the difference between nouns and verbs for some arbitrary 0. And it turns out that, for words with average sensory experience, verbs are *more* iconic than nouns. You have changed the representation of the model so that the POSVerb coefficient is more interpretable. When dealing with interactions and you are uncertain about whether you should or should not center, I recommend the motto 'If in doubt, center'.

8.3. Categorical * Categorical Interactions

This section explores interactions between two categorical variables in the context of an experimental study conducted by Winter and Matlock (2013). This study followed a 2 x 2 ('two by two') experimental design, which means there were two categorical predictors, each of which had two levels. We were interested in people's associations between conceptual similarity and physical proximity as revealed through such metaphorical expressions as *Their views on this issue are far apart* (see also Casasanto, 2008; Boot & Pecher, 2010). In one of the experiments, participants were asked to read the following text:

> The city of Swaneplam has just finished its annual budget, and so has the city of Scaneplave. Swaneplam decided to invest more in education and public healthcare this year. It will also contribute generously to its public transportation system. Similarly, Scaneplave will increase funding for education and healthcare. Also like Swaneplam, Scaneplave will dramatically expand funds for transportation this year.

Notice that the text repeatedly emphasizes the similarity between the two cities. In another condition of our experiment, dissimilarities between the two cities were highlighted. After reading either the similar or different text, participants were asked to draw two Xs on a map of an island. The distance between the two Xs was our main response variable. We reasoned that when the cities were described as being similar, participants would place the cities closer to each other on the map. In addition to this manipulation of what we call 'semantic similarity', we added a 'phonological similarity' condition. In the phonologically similar condition, the names of the two cities were *Swaneplam* and *Scaneplave*; in the phonologically different condition, the names of the two cities sounded more different from each other: *Swaneplam* and *Mouchdalt*.

Given our research questions, we want to model the distance between the two cities as a function of phonological similarity and semantic similarity. Moreover, it is plausible that these two predictors interact. To assess this, let's start by loading in the data into your current R session.

```
sim <- read_csv('winter_matlock_2013_similarity.csv')

sim
```

```
# A tibble: 364 x 3
    Sem       Phon      Distance
    <chr>     <chr>        <int>
 1 Different Similar          76
 2 Different Different       110
 3 Similar   Similar         214
 4 Different Different        41
 5 Different Different        78
 6 Different Similar          87
 7 Similar   Different        49
 8 Different Similar          72
 9 Similar   Different       135
10 Different Similar          78
# ... with 354 more rows
```

Each row in this tibble represents data from one participant. The experiment was entirely 'between-participants', that is, each participant was exposed to just one condition. The Sem and Phon columns contain information about semantic similarity and phonological similarity, which are the predictors. The Distance column contains the response, the distance between the two cities measured in millimeters.

Let us use count() to check the number of data points per condition.

```
sim %>% count(Phon, Sem)
```

```
# A tibble: 4 x 3
# Groups:   Phon, Sem   [4]
  Phon      Sem              n
  <chr>     <chr>        <int>
1 Different Different       91
2 Different Similar         86
3 Similar   Different       97
4 Similar   Similar         90
```

Finally, what about the `Distance` values? I happen to know that there is at least one missing value, which can be assessed as follows using the `is.na()` function (see Chapter 2.6). The result is a logical vector containing `TRUE` values (for NAs) and `FALSE` values for complete cases. When the `sum()` function is used on a logical vector, `TRUE` values are treated as 1s, and `FALSE` values as 0s.

```
sum(is.na(sim$Distance))
```

```
[1] 1
```

Thus, there is only one missing value. Either one of the following two `filter()` commands exclude this data point.

```
sim <- filter(sim, !is.na(Distance))

# Same as:

sim <- filter(sim, complete.cases(Distance))
```

Let's verify that the new tibble has indeed one row less:

```
nrow(sim)
```

```
[1] 363
```

Yes. To get a feel for the `Distance` measure, compute the range.

```
range(sim$Distance)
```

```
[1]   3 214
```

So, participants drew the cities anywhere from 3mm to 214mm apart from each other. Let's fit the model, first without any interactions:

```
sim_mdl <- lm(Distance ~ Phon + Sem, data = sim)

tidy(sim_mdl) %>% select(term, estimate)
```

```
         term      estimate
1 (Intercept)    79.555653
2 PhonSimilar     5.794773
3  SemSimilar   -10.183661
```

Because 'd' comes before 's' in the alphabet, R will assign the reference level to 'Different' for both predictors. This results in the slopes expressing the change from 'Different' to 'Similar'. As always, it's good to ask yourself what's in the intercept. Here, the intercept represents the estimated distance (79.6mm) for the phonologically different *and* semantically different condition (i.e., when both predictors are 0). Correspondingly, the two coefficients `PhonSimilar` and

`SemSimilar` represent the change with respect to this reference level. Because there are no interactions in this model, each of these differences can be considered in isolation. The model predicts distance to decrease by 10.2mm for the semantically similar condition. It predicts distance to increase by 5.8mm for the phonologically similar condition.

Now, fit a model with the interaction term:

```
sim_mdl_int <- lm(Distance ~ Phon * Sem, data = sim)

tidy(sim_mdl_int) %>% select(term, estimate)
```

```
                      term      estimate
1              (Intercept)     78.400000
2              PhonSimilar      8.022680
3               SemSimilar     -7.818605
4  PhonSimilar:SemSimilar     -4.592965
```

Again, the presence of an interaction should invite you to slow down and think closely about the meaning of each coefficient. It is absolutely essential that you remind yourself that you cannot interpret the effects of the two predictors in isolation anymore.

First, what's in the intercept? Well, you haven't changed the reference levels, so the intercept is still the phonologically different and semantically different condition. Next, remember that the `PhonSimilar` and `SemSimilar` coefficients are with respect to this intercept. However, because there now is an interaction in the model, the `PhonSimilar` coefficient (+8mm) is *not* the average effect of phonological similarity anymore. Instead, `PhonSimilar` is the difference between the phonologically similar and different conditions *for semantically different words only*. Likewise, the `SemSimilar` coefficient describes the effect of semantic similarity *for phonologically different words only*. The `PhonSimilar` and `SemSimilar` coefficients in the above table are called 'simple effects', a term used to describe the influence of one predictor for *a specific level of the other predictor*. Usually, researchers are interested in 'main effects', which is the average effect of one predictor, regardless of the levels of the other predictor. Misinterpreting simple effects as main effects is a huge issue in linguistics (see also Levy, 2018).

What about the interaction? The coefficient of the `PhonSimilar:SemSimilar` term is –4.6mm. This represents the change that is applied only to the phonologically similar *and* semantically similar cell. When dealing with interactions in a 2 x 2 design, I find it incredibly useful to draw a 2 x 2 table and fill in the coefficient estimates by hand. Doing this makes some things immediately apparent, so let's have a look at Table 8.1. First, notice that all cells include the intercept. Next, notice that the `PhonSimilar` coefficient (+8.0) is applied to the entire column that is 'phonologically similar', and the same goes for the `SemSimilar` coefficient (–7.8), which is applied to the entire 'semantically similar' row. However, these are not the average row-wise or column-wise differences, as one also has to include the interaction term (highlighted in bold) in calculating these averages. The table margins show row-wise and column-wise averages. This reveals that the average effect of semantic similarity is actually +10.1, not +8.0 as in the coefficient table above. Similarly, the average phonological similarity effect is –5.7 once the interaction has been taken into account, not –7.8.

If you are unsure in doing this arithmetic yourself, use predict() instead. Remember that the first step of using predict() is to create a dataset for which predictions should be generated. The code below defines two vectors: one Phon with the phonological conditions, one Sem with the semantic conditions. However, you need to make sure that you get all combinations of 'similar' and 'different' values across these two conditions. To do this, the rep() function comes in handy. It *repeats* a vector (hence the name). If you specify the argument each = 2, each element of a vector will be repeated. If you specify the argument times = 2, the entire vector will be repeated. Thus, the difference between these two arguments results in a different order in which things are repeated. For the present purposes, this is useful because it allows creating all condition combinations.

```
# Create 'different, different, similar, similar':

Phon <- rep(c('Different', 'Similar'), each = 2)

Phon
```

```
[1] "Different" "Different" "Similar"  "Similar"
```

```
# Create 'different, similar, different, similar':

Sem <- rep(c('Different', 'Similar'), times = 2)

Sem
```

```
[1] "Different" "Similar"  "Different" "Similar"
```

Once you have these two vectors, put them both into a tibble.

```
newdata <- tibble(Phon, Sem)

newdata
```

```
# A tibble: 4 x 2
  Phon      Sem
```

Table 8.1. Coefficients of the model with interaction term; the table margins show the row-wise and column-wise averages

	Phonologically different	*Phonologically similar*	
Semantically different	78.4	78.4 + 8.0	*M* = **82.4**
Semantically similar	78.4 + (–7.8)	78.4 + 8.0 + (–7.8)+(–4.6)	*M* = **72.3**
	column M = **74.5**	*column M* = **80.2**	

```
   <chr>      <chr>
1 Different Different
2 Different Similar
3 Similar   Different
4 Similar   Similar
```

This tibble has two columns, one for each of the experimental manipulations. Together, the four rows exhaust the possible combinations of conditions. This tibble is then supplied to the predict() function, together with the corresponding model sim_mdl_int.

```
# Append predictions to tibble:

newdata$fit <- predict(sim_mdl_int, newdata)

newdata
```

```
# A tibble: 4 x 3
  Phon      Sem            fit
  <chr>     <chr>        <dbl>
1 Different Different     78.4
2 Different Similar       70.6
3 Similar   Different     86.4
4 Similar   Similar       74.0
```

These predictions can then be used to compute averages. For example, what is the average difference between the semantically similar and semantically different conditions?

```
newdata %>% group_by(Sem) %>%
  summarize(distM = mean(fit))
```

```
# A tibble: 2 x 2
  Sem        distM
  <chr>      <dbl>
1 Different   82.4
2 Similar     72.3
```

Thus, the model predicts that, in the semantically similar condition, participants drew the cities 10mm closer to each other.

The other strategy to aid in the interpretation of main effects in the presence of interactions is to change the coding scheme from treatment coding (R's default) to sum coding. Remember that treatment coding assigns the values 0 and 1 to categories, whereas sum coding uses −1 and +1 (see Chapter 7). Conceptually, you can think of sum coding here as 'centering' your categorical predictors. The most important feature of sum coding for our purposes is that 0 is not any one of the categories anymore, so that the changes are expressed as differences to that specific category. Instead, sum codes have 0 halfway in between the two categories. This is best seen in action. The following code uses mutate() to define two new columns with factor versions of the condition variables, `Phon_sum` and `Sem_sum`. After this, the coding scheme of these factor vectors is changed to sum coding factor vectors is changed to sum coding factor vectors is changed to sum coding via `contr.sum()`.

```
# Convert predictors to factors:

sim <- mutate(sim,
              Phon_sum = factor(Phon),
              Sem_sum = factor(Sem))

# Change contrast coding scheme to sum coding:

contrasts(sim$Phon_sum) <- contr.sum(2)
contrasts(sim$Sem_sum) <- contr.sum(2)
```

Now, refit the model with the new sum-coded factor variables:

```
# Refit model with sum-coded predictors:

sum_mdl <- lm(Distance ~ Phon_sum * Sem_sum, data = sim)

tidy(sum_mdl) %>% select(term, estimate)
```

```
                 term    estimate
1         (Intercept)   77.353797
2           Phon_sum1   -2.863099
3            Sem_sum1    5.057543
4  Phon_sum1:Sem_sum1   -1.148241
```

What's the meaning of the `Sem_sum1` coefficient? This is *half* the difference between the two semantic conditions. Multiplying this by 2 yields about 10mm, which is the average difference between semantically different and semantically similar drawings. Why is this only *half* the average difference? This has to do with the fact that, in sum coding, changing from one category (−1) to another (+1) is a bigger change, namely 2 rather than 1 (as is the case with treatment coding). See Chapter 7 for a reminder. Either way, what is important to us is that, once the categorical predictors have been sum-coded, the coefficient table lists 'main effects' rather than 'simple effects'.[2]

What's often confusing about sum coding is the sign of the coefficient. First of all, I happily admit that for getting a quick grasp of what's going on in a sum-coded model I ignore the sign of the main effects. This can be done because if everything 'starts in the middle' it doesn't matter whether you're moving 'up' or 'down' by half a mean difference. In my own practice, I mostly use treatment coding, because I find it easier to deal with the arithmetic (since terms drop out when multiplied by 0). I only move

---

2 Given that all four cells of the design contain approximately the same amount of data, the intercept is now also approximately the grand mean (the overall mean of the response), which has a straightforward interpretation in this case: it is the average distance between the cities.

to sum coding when I want to report main effects in a paper. As long as you know how to interpret things, you may use either one of the two coding schemes—whatever is easiest for you to wrap your head around. That said, it is important to be specific about the coding scheme in the write-up your results—only this way will your audience be able to interpret the coefficients appropriately. If you are worried about interpreting coding schemes (which takes time and practice), you always have `predict()` at your disposal.

8.4. Continuous * Continuous Interactions

The final case to consider is an interaction between two continuous predictors. For this, you are going to retrace the first steps of an analysis performed by Sidhu and Pexman (2018). These researchers looked at the effect of sensory experience on iconicity (see also Chapter 6). However, they additionally considered the role of what psycholinguists call 'semantic neighborhood density'. This term describes the idea that there are certain regions in your mental lexicon that are quite 'crowded' or 'dense', with lots of words that are connected to each other by virtue of having similar meanings. It has been proposed that iconicity may lead to confusion, because, with many iconic forms, similar meanings will also sound similar (Gasser, 2004; Christiansen & Chater, 2016). Sidhu and Pexman (2018) reasoned that language should be biased against iconicity specifically in semantically dense neighborhoods, where there is more room for confusion. In sparse neighborhoods, iconicity is not as dangerous, as there is less opportunity to confuse concepts.

Let's load the data from Sidhu and Pexman (2018) into your current R session. In line with the snazzy title of their paper (advance-published in 2017), 'Lonely sensational icons', let's call the tibble `lonely`:

```
lonely <- read_csv('sidhu&pexman_2017_iconicity.csv')

lonely
```

```
# A tibble: 1,389 x 4
   Word    SER   ARC Iconicity
   <chr> <dbl> <dbl>     <dbl>
 1 one    1.55 0.702     1.85
 2 him    2.55 0.689     0.583
 3 she    1.60 0.687     0.714
 4 me     2.33 0.664     0.600
 5 he     1.40 0.694     1.06
 6 mine   2.08 0.641     1.50
 7 near   2.10 0.674     0.538
 8 spite  2.91 0.625     2.86
 9 few    1.55 0.697     2.50
10 none   1.73 0.661     0.833
# ... with 1,379 more rows
```

The tibble contains four columns. The `Word`, `SER`, and `Iconicity` columns are already familiar to you from this chapter and Chapter 6. The `ARC` column stands for 'average radius of co-occurrence'. I won't go into detail about how this measure is computed—all you need to know is that this is a measure of semantic neighborhood density taken from Shaoul and Westbury (2010). Small ARC values indicate sparse semantic neighborhoods (less potential for confusion); large ARC values indicate dense neighborhoods (more potential for confusion). Given that iconicity can lead to confusion, words associated with large ARC values (in dense neighborhoods) should be less iconic.

Sidhu and Pexman (2018) decided to exclude words with low iconicity ratings.[3]

```
lonely <- filter(lonely, Iconicity >= 0)
```

Let us fit a model with SER and ARC as predictors, including their interaction.

```
lonely_mdl <- lm(Iconicity ~ SER * ARC, data = lonely)

tidy(lonely_mdl) %>% select(term, estimate)
```

```
          term     estimate
1  (Intercept)   1.3601014
2          SER   0.3612026
3          ARC  -0.7929281
4      SER:ARC  -0.5255308
```

It is important to remember that each coefficient is shown for 0 of the other variables. The intercept is the predicted iconicity rating of a word with 0 sensory experience and 0 semantic neighborhood density (ARC). The slope of the ARC effect (−0.79) is shown for a sensory experience rating of 0. Likewise, the slope of the SER effect is shown for an ARC of 0.

This model is hard to interpret. First, the fact that the continuous predictors aren't centered means that the SER and ARC effects are reported for some arbitrary 0. It would be much nicer to report the ARC effect for words with average sensory experience and, likewise, to report the SER effect for words with average semantic neighborhood density. A second interpretational difficulty arises from the fact that the magnitude of the SER and ARC slopes are difficult to compare as each variable has a different metric. That is, a one-unit change is a different step-size for each variable. Let's standardize both predictors (remember from Chapter 5 that standardized variables are also centered variables). The following `mutate()` command standardizes both continuous predictors.

---

3 The lower end of the iconicity rating scale may lack what is called 'construct validity'. It is not clear at all that 'negative' iconicity was interpreted consistently by our participants. In their paper, Sidhu and Pexman (2018) used a more complex decision procedure to exclude words with low iconicity.

```
# Standardize continuous predictors:

lonely <- mutate(lonely,
                 SER_z = (SER - mean(SER)) / sd(SER),
                 ARC_z = (ARC - mean(ARC)) / sd(ARC))
lonely
```

```
# A tibble: 1,389 x 6
   Word    SER   ARC Iconicity   SER_z   ARC_z
   <chr> <dbl> <dbl>     <dbl>   <dbl>   <dbl>
 1 one    1.55 0.702      1.85   -1.74    1.16
 2 him    2.55 0.689     0.583  -0.745    1.06
 3 she    1.6  0.687     0.714   -1.69    1.05
 4 me     2.33 0.664     0.6    -0.956   0.871
 5 he     1.4  0.694      1.06   -1.89    1.10
 6 mine   2.08 0.641      1.5    -1.21   0.691
 7 near   2.1  0.674     0.538   -1.19   0.949
 8 spite  2.91 0.625      2.86  -0.382   0.567
 9 few    1.55 0.697       2.5   -1.74    1.13
10 none   1.73 0.661     0.833   -1.56   0.842
# ... with 1,379 more rows
```

Now that you have standardized predictors, refit the model. Let's call this new model `lonely_mdl_z`.

```
# Fit model with standardized predictors:

lonely_mdl_z <- lm(Iconicity ~ SER_z * ARC_z,
                   data = lonely)

tidy(lonely_mdl_z) %>% select(term, estimate)
```

```
          term      estimate
1 (Intercept)    1.15564895
2        SER_z    0.07115308
3        ARC_z   -0.32426472
4  SER_z:ARC_z   -0.06775347
```

In this model, the `ARC_z` and `SER_z` coefficients are shown for each other's respective averages, because the meaning of 0 has changed for both variables. After standardization, 0 SER is now the mean sensory experience rating; 0 ARC is now the mean semantic neighborhood density. Let's first interpret the `ARC_z` and `SER_z` coefficients before dealing with the interaction. First, the slope of ARC is negative. This means that, in denser neighborhoods (higher values of ARC), words are indeed less iconic, as predicted by Sidhu and Pexman (2018). In addition, the slope of `SER_z` is positive, confirming the idea that perceptual words are generally more iconic (Winter et al., 2017).

Moving on to the interaction, notice that the slope for this term is negative (`SER_z:ARC_z`: –0.07). This can be read as follows: when `SER_z` and `ARC_z` *both* increase, words are actually predicted to be *less* iconic. You can think of this as the two effects cancelling each other out. It also helps to think about one variable at a time, considering what the interaction means for that variable. For example: since the SER_z slope is otherwise positive, but the interaction SER_z:ARC_z is negative, the SER effect is diminished for high levels of the ARC variable. Put differently: in dense semantic neighborhoods, there is less of a sensory experience effect.

Sidhu and Pexman (2018) opted to visualize the interaction via a partial plot where the relationship between SER and iconicity is shown for representative values of ARC. They chose −1.5 standard deviations and +1.5 standard deviations above the ARC mean as the value to plot the Iconicity ~ SER relationship for. A version of their plot is shown in Figure 8.4a.

Figure 8.4a shows that, for high levels of ARC (dense semantic neighborhoods), the relationship between sensory experience and iconicity is basically nil (a flat line). For semantic neighborhoods of intermediate density (ARC = 0), there is a very weak positive relationship. The SER effect is more strongly positive in sparse semantic neighborhoods (ARC = −1.5).

While Figure 8.4a is certainly useful, notice that it is a very partial representation of what's actually going on with this model. The figure shows only three lines for three different arbitrarily chosen values of the ARC predictor. However, the model allows you to make predictions for a whole series of ARC values.

Figure 8.4b shows a 3D version of the same relationship. This will take some time to digest, so let me walk you through it. It's important to realize that with two continuous predictors you can span a two-dimensional grid of predictor values; for example, ARC = 0, 1, 2, and then SER = 0, 1, 2 for every ARC value, and so on. Regression makes predictions for all possible combinations of the predictor values. These predictions can be represented in terms of height, extending upwards into a third dimension. The resulting plane is called the regression plane.

To make sense of a 3D plot such as the one shown in Figure 8.4b, it is often useful to consider each axis in isolation first. Focus your eye on the SER predictor. When looking at what is labeled the 'bottom edge' in Figure 8.4b, there is not a particularly

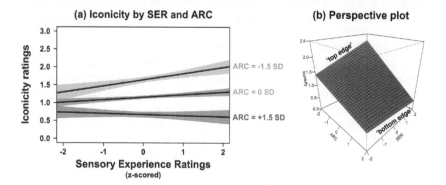

Figure 8.4. (a) Interaction plot showing the relationship between sensory experience and for different levels of semantic neighborhood density ('ARC'); (b) predicted regression plane of the relationship between iconicity, ARC and SER; the z-axis (height) shows iconicity (up = more iconic, down = more arbitrary); the x and y axes show the continuous predictors, SER and ARC

(a) No interaction   **(b) Interaction**

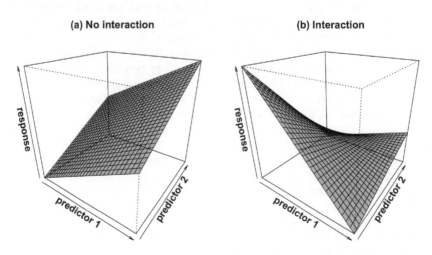

Figure 8.5. (a) A straight regression plane *without* an interaction and (b) a twisted regression plane *with* an interaction; notice that (b) only shows one possible interaction out of many possible interactions

strong relationship between SER and iconicity. This is the regression line for semantically dense neighborhoods (higher ARC values). When looking at the 'top edge', there is a positive relationship between SER and iconicity. Thus, depending on what value ARC assumes, the slope of the SER effect changes. This actually means that the regression plane seen in Figure 8.4b is twisted.

All of this 3D stuff is actually quite useful for conceptualizing interactions between continuous predictors. Figure 8.5a shows that when there is *no* interaction, the plane is straight. Figure 8.5b shows one of many possible interactions between two continuous predictors. Notice that in this case the plane isn't straight anymore, it's twisted. This is the conceptual analogue to what you've seen in the case of categorical * continuous interactions, where parallel lines indicated the absence of an interaction, and non-parallel lines indicated the presence of an interaction.

The particular interaction shown in Figure 8.5b is of an interesting nature. For low values of predictor 2, there is a negative relationship between predictor 1 and the response. For high values of predictor 2, there is a positive relationship between predictor 1 and the response. Thus, the slope of predictor 1 reverses sign as you move along predictor 2. There are many situations in linguistics where the coefficient of a continuous predictor reverses sign depending on another continuous predictor. For example, in Perry et al. (2017), we found a positive relationship between iconicity and word frequency for young children, but a negative relationship between iconicity and word frequency for older children. Thus, the relationship between frequency and iconicity was modulated by the continuous predictor age.

8.5. Nonlinear Effects

In the context of talking about multiplicative effects, it also makes sense to talk about situations in which you have to deal with nonlinear effects. What if the data does not follow a straight line? Like interactions, such nonlinearities can be modeled by multiplications of predictor variables.

In a paper that won them the Yelp Dataset Challenge Award, Vinson and Dale (2014) looked at how the information density of messages is influenced by a language user's affective state, as operationalized through whether a restaurant's review on the Yelp app is overall positive, negative, or neutral. Vinson and Dale (2014) used several measures for computing the information density of Yelp reviews. Here, we will only look at one of these measures, the average conditional information (ACI), which quantifies how unexpected a word is given the word that immediately preceded it. In information theory, 'unexpectedness' is a measure of how informative something is. Vinson and Dale (2014) correlated this measure with the user's rating of each review (Yelp uses a five-star system).

The dataset I make available to you is a small subset of the much larger analysis conducted by Vinson and Dale (2014). Each of the 10,000 rows in this tibble represents one review. The `across_uni_info` column is the average ACI over the entire review. Higher values indicate more informative reviews.

```
vinson_yelp <- read_csv('vinson_dale_2014_yelp.csv')

vinson_yelp
```

```
# A tibble: 10,000 x 2
   stars across_uni_info
   <int>           <dbl>
 1     2            9.20
 2     1           10.2
 3     4            9.31
 4     3            9.27
 5     3            9.34
 6     4            9.68
 7     4            9.79
 8     4            9.47
 9     3           10.6
10     5            9.23
# ... with 9,990 more rows
```

Let us compute averages of the information density measure as a function of the Yelp review stars. This is achieved with the following code. In this pipeline, the `vinson_yelp` tibble is first piped to the `group_by()` function, which groups the tibble by stars. The grouping means that the following `summarize()` function computes averages per review star. These averages (`AUI_mean`) are then plotted against the review stars to yield Figure 8.6.

```
vinson_yelp %>% group_by(stars) %>%
  summarize(AUI_mean = mean(across_uni_info)) %>%
  ggplot(aes(x = stars, y = AUI_mean)) +
  geom_line(linetype = 2) +
  geom_point(size = 3) +
  theme_minimal()
```

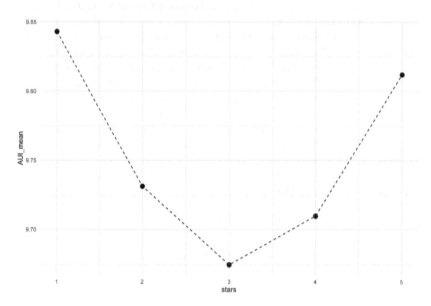

Figure 8.6. Average information density as a function of restaurant review ratings (Yelp); the analysis is based on a subset of the data from Vinson and Dale (2014)

The pattern shown in Figure 8.6 suggests that reviews are more informative if they are less neutral. Or, to put it plainly, more strongly negative or more strongly positive reviews contain more unpredictable words.

To model such a curve with regression, polynomial effects can be incorporated into the model. In this case, visual inspection of the averages suggests that there is a quadratic ('parabolic') effect. The parabola of a variable x is the function $f(x) = x^2$, which is just x multiplied by itself ($x*x$). The following plot demonstrates this (see Figure 8.7, left plot). The sequence of numbers $-10:10$ is stored in the vector x. This vector is then multiplied by itself via power notation ^2.

```
x <- -10:10
plot(x, x ^ 2, type = 'b')
```

Similarly, multiplying a sequence of numbers three times by itself ($x*x*x$) results in the characteristic S-shaped pattern of cubic curves (see Figure 8.7, right plot).

```
x <- -10:10
plot(x, x ^ 3, type = 'b')
```

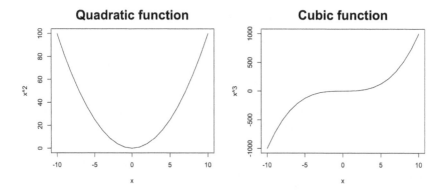

Figure 8.7. Quadratic and cubic functions of x

The trick in polynomial regression is to enter a polynomially transformed version of a predictor. For example, in E8.3, a quadratic version of the predictor x is entered into the model. Regression then estimates a coefficient for this quadratic variable based on the data. The size of the coefficient for this variable quantifies the degree to which the relationship between x and y is parabolic. In E8.3, the coefficient for the quadratic effect is β_2. When β_2 is estimated to be large and positive, the model will predict a U-shaped curve; when β_2 is estimated to be large and negative, the model will predict an inverted U-shaped curve. The more β_2 tends towards 0, the more everything falls on a straight line.

$$y = \beta_0 + \beta_1 x + \beta_2 x^2 \tag{E8.3}$$

In other words: you specify the polynomial form you predict (in this case, a quadratic), and the data determines the coefficient estimates that your model predicts based on this specification. If there is not actually a quadratic or cubic pattern in the data, the corresponding coefficient estimates will be very small.

So, to do this for Vinson and Dale's (2014) Yelp review data, center the star ratings and create a squared version of this, which is just the star ratings variable multiplied by itself.

```
# Center and square star ratings:

vinson_yelp <- mutate(vinson_yelp,
                      stars_c = stars - mean(stars),
                      stars_c2 = stars_c ^ 2)
```

The model then includes the linear and the quadratic star ratings as predictors of average information density.

```
yelp_mdl <- lm(across_uni_info ~ stars_c + stars_c2,
               data = vinson_yelp)
```

```
tidy(yelp_mdl) %>% select(term:estimate)
```

```
          term     estimate
1 (Intercept)   9.69128114
2       stars_c  0.04290790
3      stars_c2  0.03736348
```

There is a positive coefficient for the quadratic effect. To interpret this model, it's easiest to compute predictions. The following code uses the familiar `predict()` function to compute the fitted values for each of the ratings.

```
# Create tibble for predict():

yelp_preds <- tibble(stars_c =
                        sort(unique(vinson_yelp$stars_c)))

# Square star ratings:

yelp_preds <- mutate(yelp_preds, stars_c2 = stars_c ^ 2)

# Append model fit:

yelp_preds$fit <- predict(yelp_mdl, yelp_preds)

yelp_preds
```

```
# A tibble: 5 x 3
   stars_c  stars_c2      fit
     <dbl>     <dbl>    <dbl>
1    -2.68      7.21     9.85
2    -1.68      2.84     9.73
3   -0.685     0.469     9.68
4    0.315    0.0992     9.71
5     1.32      1.73     9.81
```

Notice how the numbers in the `fit` column first become smaller, and then larger again. Based on these fitted values, you can produce a plot of the predictions, which will look almost exactly the same as Figure 8.6. The predictions follow a clear U-shaped pattern.

```
yelp_preds %>%
  ggplot(aes(x = stars_c, y = fit)) +
  geom_point(size = 3) +
  geom_line(linetype = 2) +
     theme_minimal()
```

It is possible to add additional polynomials to model more complex curves. For example, a model could take the form 'y ~ x_1 + x_2 + x_3 + x_4', where x_2 is the quadratic effect ($x*x$), x_3 is the cubic effect ($x*x*x$), x_4 is the quartic effect ($x*x*x*x$), and so on. Adding such 'higher-order' polynomials means that increasingly wiggly curves can be modeled with increasing fidelity. However, be careful with this approach, as higher-order polynomials are often difficult to interpret. Moreover, you need to keep in mind that polynomials are just mathematical objects, and especially higher-order polynomials may not relate in a meaningful fashion to any linguistic theories. In my own practice, I very often model quadratic effects because these often relate to specific theoretical mechanisms. For example, in memory research, there are 'primacy effects' and 'recency effects' which make people remember the beginning or ends of lists better and result in a U-shaped curve of memorization performance across the list. This could easily be modeled with a quadratically transformed version of a 'list position' predictor. In Hassemer and Winter (2016) and Littlemore et al. (2018), we used polynomials to capture the fact that some effects have plateaus for high values.

If you are purely interested in modeling nonlinear trajectories with high fidelity, you may consider using generalized additive models (GAMs). These are an extension of regression that are much less restricted than polynomial regression. A coverage of GAMs is beyond the scope of this book, but the concepts learned here prepare you well for fitting GAM models. Luckily, there is a wealth of tutorials on GAMs available (Winter & Wieling, 2016; Sóskuthy, 2017; Wieling, 2018).

8.6. Higher-Order Interactions

So far, this chapter has covered interactions between two variables. It's also possible to fit interactions between more than two variables. In some cases, this may be theoretically motivated. However, the interpretational problems inherent in models with interactions become amplified for more complex interactions. Let's say you had three predictors, A, B, and C, which you entered into an interaction (A * B * C). In that case, all *three* predictors cannot be interpreted in isolation anymore. Moreover, the two-way interactions between each pair of the predictors (A:B, B:C, and A:C) is modulated by the three-way interaction!

In some cases, three-way interactions may be theoretically motivated. Sometimes, even four-way interactions make sense and may be predicted. However, in general, it's a good strategy to avoid fitting models with very complex interaction terms—especially if you are new to statistical modeling. Focus on those interactions for which there is a clear theoretical rationale.

You can also use the model formulas strategically to limit the number of interactions. Have a look at the following notation:

```
y ~ (A + B) * C
```

This model fits the two-way interactions between A and C, and between B and C, but it doesn't fit the three-way interaction, and it also doesn't fit the interaction between A and B.

In this context, it's also good to remember that the star symbol '*' is just a shorthand, and that one can specify each individual interaction term separately using the colon ':'. The model formula above corresponds to the following equivalent regression specification:

```
y ~ A + B + A:C + B:C
```

8.7. Chapter Conclusions

This chapter has introduced you to interactions, which are estimated via the coefficients of multiplicative predictors. You can think of an interaction as a situation in which two predictors are 'interlocked', having a conjoined influence on the response. If your model features an interaction, you cannot interpret the predictors that participate in the interaction in isolation anymore. In the presence of an interaction, the effects of one predictor depend on another predictor.

I have walked you through three types of interactions: continuous * categorical, categorical * categorical, and continuous * continuous. Throughout these examples, you have seen that recoding your predictor (such as centering a continuous variable or sum-coding a categorical predictor) may aid interpretation. In addition, you should extensively use the `predict()` function to understand your model, and you should compare your model coefficients to visualizations of the data, or to descriptive averages.

Finally, I have shown how to fit polynomials to model nonlinear data. Altogether, this chapter has massively increased the expressive potential of your statistical modeling, allowing you to model more complex datasets, as well as more complex theories.

8.8. Exercises

8.8.1. Exercise 1: Interactions with Multi-Level Categorical Predictors

Go back to the step where you filtered the `icon` tibble to create `NV`, the tibble that only included the data for nouns and verbs. Create a new subset that also includes adjectives, so that the `POS` column now has three categories. Refit the model with the part-of-speech interaction. How many interaction coefficients are there now?

8.8.2. Exercise 2: Hand-Coding Categorical Predictors with Interactions

To reinforce coding schemes, recreate the dummy codes for the similarity model discussed in this chapter. Doing this by hand is generally not recommended, since R does it anyway, but it's useful for pedagogical reasons as it helps to reinforce what's going on 'behind the scenes' when R fits an interaction between two categorical variables. The following code uses the `ifelse()` function to create two new columns. These are then multiplied to create the interaction term `int01`.

```
sim <- mutate(sim,
              phon01 = ifelse(Phon == 'Similar', 1, 0),
              sem01 = ifelse(Sem == 'Similar', 1, 0),
              int01 = phon01 * sem01)

tidy(lm(Distance ~ phon01 + sem01 + int01, data = sim))
# output not shown
```

Compare the resulting output to the treatment-coded model discussed in the chapter.

9 Inferential Statistics 1

Significance Testing

9.1. Introduction

This chapter is the first one to explicitly deal with inferential statistics. As briefly mentioned in Chapter 3, when doing inferential statistics, *sample estimates* are used to estimate and make inferences about *population parameters*. Most of the time, researchers work with limited samples. Each sample is a 'snapshot' of the population, and randomly drawing different samples from the population means that each sample yields slightly different estimates. Thus, the process of making inferences on the population is always uncertain. Gelman and Hill (2007: 457) say that uncertainty "reflects lack of complete knowledge about a parameter". Inferential statistics quantifies this uncertainty so that you can express how confident you are in making claims about the population.

Figure 9.1 is a schematization of the inferential process. Chapter 3 already talked about the fact that the sample mean $\bar{x}$ is an estimate of the population mean μ, and that the sample standard deviation s is an estimate of the population standard deviation σ. In the context of regression, each regression coefficient b is an estimate of the corresponding regression coefficient in the population, β (pronounced 'beta').

What counts as a 'population' in linguistics is, in fact, a non-trivial question. In sociolinguistic and psycholinguistic studies, one population of interest is often the set of all speakers of a particular linguistic variety. For example, you may have collected speech data from a sample of 90 native speakers of the 'Brummie' accent, a variety of British English spoken in the city of Birmingham (UK). This sample of 90 speakers vastly undersamples the population of interest (Birmingham is the second biggest city in the UK). Nevertheless, you want to assess how likely it is that our results say something about *all* Brummie speakers.

For many linguistic applications, the population of interest is not a population of speakers, but a population of linguistic items. In fact, in most of my own work (corpus analysis and lexical statistics) I sample primarily from populations of linguistic words, rather than from populations of speakers. In this case, the set of all English words is the relevant population that we want to make inferences upon. One also has to think about such a 'population of items' in the context of experimental studies, which almost always sample not only participants but also items.

The big question is: how can you make inferences about the population? In fact, there are different ways of answering this question, with different schools of thought

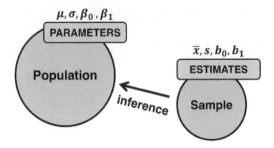

Figure 9.1. Schematic depiction of a sample drawn from a specified population; sample estimates (Roman letters) are used to make inferences about population parameters (Greek letters)

that are philosophically quite distinct from each other (for a great discussion, see Dienes, 2008). Perhaps the most widely used approach is null hypothesis significance testing, often abbreviated NHST.

NHST is ubiquitous, but from the outset it should be said that it has also received much criticism (e.g., Goodman, 1999; Krantz, 1999; Nickerson, 2000; Sterne & Smith, 2001; Gigerenzer, 2004; Kline, 2004; Thompson, 2004; Hubbard & Lindsay, 2008; Cumming, 2012, 2014; Morey, Hoekstra, Rouder, Lee, & Wagenmakers, 2016; Vasishth & Gelman, 2017, among many others). I think it is dangerous for statistics textbooks to teach significance testing without mentioning that there is controversy surrounding this approach. I will discuss some of these issues later.

NHST, as it is used today, has grown over time out of several distinct historical strands. It incorporates ideas from Sir Ronald Fisher, Jerzy Neyman, Egon Pearson, and others. Perezgonzalez (2015) reviews how what is considered significance testing today is actually a logically inconsistent conglomerate of the Fisherian tradition of significance testing and the Neyman-Pearson approach. Although a full disentanglement of the different strands of significance testing is beyond the scope of this book (see Perezgonzalez, 2015), I will try my best to convey significance testing in such a way that certain common thinking traps are avoided. In addition, following the recommendations of Cumming (2012, 2014), I will try to alleviate at least some of the concerns surrounding NHST by placing stronger emphasis on effect sizes (this chapter) and interval estimates (next chapter).

At the heart of significance testing lies a particular view of probability, sometimes called 'frequentism'. Frequentists interpret probabilities in terms of long-run relative frequencies. Let us exemplify what this means in the context of predicting a coin flip. For a frequentist, the statement that the probability of heads is $p = 0.5$ cannot be meaningfully made about a singular coin flip, which comes up as either heads or tails. Instead, the frequentist thinks of the probability $p = 0.5$ as the long-run relative frequency of an infinite number of coin flips. As you tally more and more coin flips, the relative frequency converges towards 0.5. NHST is designed with such long-run frequencies in mind, with the goal of keeping the rate of making wrong claims about the population at a low level. The beauty of NHST lies in the fact that, if it is used correctly, a researcher has a well-specified error rate for a statistical test.

9.2. Effect Size: Cohen's *d*

Let us start with a concrete example, modeling a difference between two groups. Consider the fact that women and men differ in their vocal frequencies. What our ear perceives as 'voice pitch' is commonly measured as 'fundamental frequency' in the unit of 'Hertz'. Imagine you conducted a study with a sample of 100 women and 100 men, finding that their fundamental frequencies differed on average by 100 Hz. The problem is that sampling itself is a random process. Whenever you sample *any* two groups from a population of interest, you are bound to find small differences due to chance processes alone. This is the case even if there is actually no difference in the population. So, is 100 Hz a big or small difference? And is it enough to conclude that there's a difference between women and men in the population this sample was drawn from? More generally, what factors *should* affect our confidence in there being such a difference in the population?

I will frame the following discussion in terms of three crucial 'ingredients' that affect the confidence with which you make claims about the population:

1   The magnitude of a difference

All else being equal, the bigger the difference between two groups, the more you should expect there to be a difference in the population.

2   The variability in the data

All else being equal, the less variability there is within the sample, the more certain you can be that you have estimated a difference accurately.

3   The sample size

All else being equal, bigger samples allow you to measure differences more accurately.

Let's focus on the first ingredient, the magnitude of an effect. In the case of the voice pitch example, the effect has a raw magnitude of 100Hz. This is an *unstandardized* measure of effect size. All forms of inferential statistics take effect size into account to some extent and, all else being equal, stronger effect sizes result in stronger inferences. If the 100 Hz difference between women and men had been, say, 5 Hz, you would be much less confident that there is actually a difference in the population.

There are also *standardized* measures of effect size, of which you have already seen R^2 (Chapter 4 and 6), Pearson's *r* (Chapter 5), and standardized regression coefficients (Chapter 6). These standardized measures of effect size are more interpretable across studies. In the context of this discussion, it makes sense to introduce yet another standardized measure of effect size, Cohen's *d*. This statistic is used to measure the strength of a difference between two means ($\bar{x}_1$ and $\bar{x}_2$). The formula for Cohen's *d* is simply the difference between two means (the raw strength of an effect) divided by the standard deviation of both two groups together (the overall variability of the data). As you saw for other statistics in this book, the division by some measure of variability (in this case the standard deviation *s*) is what makes this a standardized measure.

$$d = \frac{\bar{x}_1 - \bar{x}_2}{s}.$$

(E9.1)

Notice that this formula actually combines two of the ingredients, namely, ingredients one and two ('magnitude of an effect', 'variability in the data'). Cohen's *d* is going to be large when either the difference between the group means is very large or when the standard deviation is very small *s*. Informally, I like to think of this as a signal-to-noise ratio. You are able to measure a strong effect either if the signal is very strong (large difference) or if the 'noise' is very weak (small standard deviation).

Cohen (1988) discussed a rule of thumb according to which "small", "medium" and "large" effects are $d = |0.2|, |0.5|$ and $|0.8|$, respectively. In mathematical notation, the vertical bars indicate absolute values, which means that the effect size can be positive or negative (both $d = 0.8$ and $d = -0.8$ are large effects). Whether *d* is positive or negative only depends on which mean is subtracted from which other mean.

In fact, another standardized measure of effect size you already know has a similar conceptual structure to Cohen's *d* namely, Pearson's correlation coefficient *r* (Chapter 5). The formula for the computation of Pearson's *r* is as follows:

$$r = \frac{s_{x,y}^2}{s_x s_y} \tag{E9.2}$$

The $s_{x,y}^2$ is the numerator (the number above the line in the fraction), it describes the 'co-variance' of *x* and *y*. As the name suggests, the co-variance measures how much two sets of numbers vary together.[1] The deminator (the divisor below the line in the fraction) contains the standard deviations of the *x* and *y* variables (s_x and s_y) multiplied by each other. The standard deviation is always given in the unit of the data. Because of this, dividing the co-variance $s_{x,y}^2$ by the product of s_x and s_y divides the metric of the data out of this statistic, yielding a dimensionless statistic that can be compared across datasets, Pearson's *r*.

And just as is the case with Cohen's *d*, I want you to realize that Pearson's correlation coefficient *r* actually combines two of the ingredients, namely, the 'magnitude of an effect' and 'the variability in the data'. Thus, conceptually, this formula actually has the same signal-to-noise structure as Cohen's *d*. If the covariation of two variables is strong (a strong 'signal'), *r* will be large. If the variability in s_x and s_y is small (little 'noise'), then *r* will be large as well. Thus, to sum up, both Pearson's *r* and Cohen's *d* grow when the numerator ('signal') is very large, or when the denominator ('noise') is very small.

Figure 9.2 serves to strengthen our intuitions regarding these ideas. For this graph, I want you to allow yourself something that you are otherwise not allowed to do, which is to pretend—for the time being—that you can make conclusions about a population from a graph alone. Think about the following question. How confident are you that the two distributions (density curves) come from the same distribution?

---

1 The formula for the covariance is $s_{x,y}^2 = \frac{1}{n-1} \sum_{i=1}^{n} (x_i - \bar{x})(y_i - \bar{y})$. The term $(x_i - \bar{x})$ measures the x-distance of each data point from the x-mean; the term $(y_i - \bar{y})$ does the same for the y-mean.

Let's illustrate how this works if two variables are strongly *positively* correlated. If a data point is much above the x-mean and *also* much above the y-mean, the multiplication of the two deviation scores makes the co-variance grow. The same goes for the case in which the x-deviations and y-deviations are both negative, since multiplying two negative numbers yields a positive number.

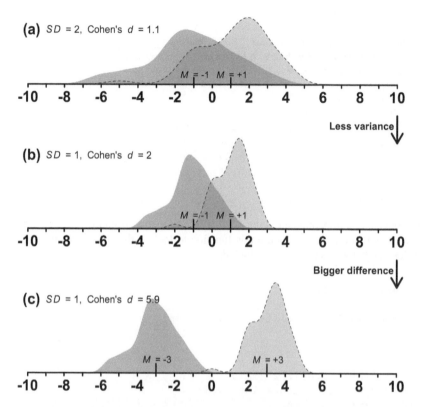

Figure 9.2. A representation of how within-group variance and magnitude of the difference between two group means impacts our ability to tell two distributions apart

Informally, then, ask yourself how easy it is to visually detect the difference between the two groups.

From Figure 9.2a to Figure 9.2b, the standard deviation is halved, which decreases the overlap between the two distributions. This makes it easier to tell the two groups apart, which shows you how, *even if the means don't change at all*, you may be more confident about the presence of a difference if the standard deviations shrink. Finally, if the group difference is then increased (Figure 9.2c), it is even easier to tell the two distributions apart. This succession of plots thus shows the interplay between the magnitude of an effect and the variability in the data. Notice how Cohen's *d* increases when the variance is decreased, and it further increases when the group differences are increased.

9.3. Cohen's d in R

Let's compute Cohen's *d* for the taste/smell dataset from Chapter 7. First, load the data from Winter (2016) back into your current R session.

```
library(tidyverse)

chem <- read_csv('winter_2016_senses_valence.csv') %>%
  filter(Modality %in% c('Taste', 'Smell'))
chem %>% print(n = 4)

# A tibble: 72 x 3
  Word       Modality     Val
  <chr>      <chr>      <dbl>
1 acidic     Taste       5.54
2 acrid      Smell       5.17
3 alcoholic  Taste       5.56
4 antiseptic Smell       5.51
# ... with 68 more row
```

Cohen's *d* is implemented in the `cohen.d()` function from the `effsize` package (Torchiano, 2016). Let's use this function to look at the strength of the valence difference between taste and smell words.

```
library(effsize)

cohen.d(Val ~ Modality, data = chem)

Cohen's d

d estimate: 1.037202 (large)
95 percent confidence interval:
      inf        sup
0.5142663 1.5601377
```

The difference between taste and smell words has a large effect size, $d = -1.04$.

9.4. Standard Errors and Confidence Intervals

Notice how the formulas for Cohen's *d* or Pearson's *r* have no term for the sample size *N*. This means that large effects can be obtained even for very small samples. This can easily be demonstrated with the following code, which creates a perfect correlation with just two data points, resulting in the highest possible correlation coefficient, $r = 1.0$.

```
# Correlate the points [1,2] and [2, 3]:

x <- c(1, 2)   # create x-values

y <- c(2, 3)   # create y-values

# Perform correlation:

cor(x, y)
```

```
[1] 1
```

If this dataset is your sample, how confident are you about this perfect correlation saying something about the population? Given that there are only two data points, any claims about the population are obviously on very shaky grounds, showing that effect size alone is clearly not enough. Standardized effect size measures such as Cohen's *d* and Pearson's *r* only contain two of the three ingredients—they are missing the sample size.

The key workhorse that incorporates N into our inferential statistics is the standard error. When estimating a population mean μ via a sample mean $\bar{x}$, the standard error is defined as follows:[2]

$$SE = \frac{s}{\sqrt{N}} \tag{E9.3}$$

This standard error combines two of the ingredients, namely the 'variability in the data' (the numerator s) and the 'sample size' (the denominator $\sqrt{N}$). Informally, the standard error is often paraphrased as indicating the precision with which a quantity is measured. Smaller standard errors measure the corresponding parameters (such as the mean) more precisely. Alternatively, large standard errors indicate that you are more uncertain in your estimates. If N is large or s is small, the standard error is small as well. In other words, you can measure a mean precisely either if there's lots of data or if there's little variation.

Standard errors can be used to compute '95% confidence intervals', often abbreviated 'CI'. As you will see, confidence intervals have a lot going for them (Cumming, 2012, 2014), although they are not without their critics (Morey et al., 2016). Confidence intervals are calculated by taking the sample estimate—in this case, the mean—and computing the boundary that is about 1.96 times the standard error above and below this sample estimate:[3]

$$CI = \left[\bar{x} - 1.96 * SE, \bar{x} + 1.96 * SE\right]. \tag{E9.4}$$

Let's exemplify this with some easy-to-work-with numbers: for a sample mean of 100 and a standard error 10, the lower bound of the 95% confidence interval is $100 - 1.96 * 10 = 80.4$. The upper bound is $100 + 1.96 * 10 = 119.6$. Thus, based on this mean and this standard error, the 95% confidence interval is $\left[80.4, 119.6\right]$.

Because the confidence interval is based on the standard error, it also depends on the same two ingredients, the standard deviation and N, as highlighted by Figure 9.3. Notice how the confidence interval shrinks when more data is added (middle dataset). Notice how the confidence interval grows again when the variance in the dataset is increased (dataset to the right of Figure 9.3).

But what, really, does the 95% confidence interval *mean*? This is where things become a bit tricky. The 95% confidence interval is enmeshed with the frequentist statistical philosophy mentioned earlier in this chapter. Remember that, according to this view of

2 The formulas for other types of standard errors (for example, for regression coefficients) are different from this one, but conceptually the overall logic stays the same.

3 I am skipping over some important technical details here. In this example, I used the value '1.96' to compute the 95% confidence interval; however, the precise value differs depending on sample size. The larger the sample size, the more this value will converge to be 1.96.

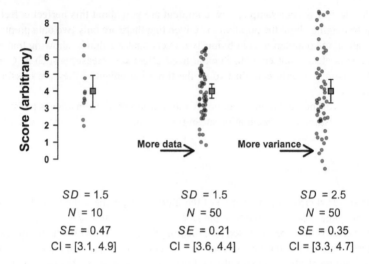

SD = 1.5	SD = 1.5	SD = 2.5
N = 10	N = 50	N = 50
SE = 0.47	SE = 0.21	SE = 0.35
CI = [3.1, 4.9]	CI = [3.6, 4.4]	CI = [3.3, 4.7]

Figure 9.3. Errors bars indicate the 95% confidence interval; the interval shrinks as the number of data points increases, and it grows again if the standard deviation increases

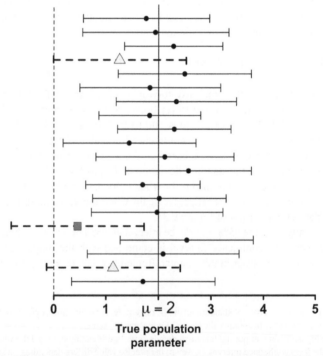

Figure 9.4. "The dance of the confidence intervals" (Cumming, 2012, 2014); see description in text

probability, objective probability cannot plausibly be assigned to singular events; a coin toss is either head or tail. Similarly, the actual population parameter of interest may or may not be inside the confidence interval—you will actually never know for sure. However, if you imagine an infinite series of experiments and compute a confidence interval each time, 95% of the time this interval would contain the true population parameter.

Figure 9.4 illustrates this idea, showing what Cumming (2012, 2014) calls "the dance of the confidence intervals". For the construction of Figure 9.4, I sampled 60 data points from the same underlying distribution each time. The distribution was defined with the population mean $\mu = 2$ and the population standard deviation $\sigma = 5$. You can think of this as multiple researchers all trying to establish where the population parameter is, with each researcher drawing a different sample.

The dashed confidence intervals with triangles include 0, and the dashed confidence interval with the square does not even cover the true population mean! Thus, by spanning confidence intervals around your sample estimates, you capture the population parameter μ *most of the time*. You can be confident in the procedure working out in the long run, but you cannot be confident about a particular dataset at hand.

9.5. Null Hypotheses

When engaging in null hypothesis significance testing, you are imposing a hard-cut decision rule to the confidence intervals shown in Figure 9.4. In other words, you straitjacket the interval nature of the confidence intervals into a procedure that leads to binary decisions, 'yes' or 'no'. While this may be conceptually appealing due to its simplicity, the practice of making binary decisions based on data is also one of the main criticisms raised against significance testing. It has been argued to lead to "dichotomous thinking" (Cumming, 2012, 2014) or "lazy thinking" (Gardner & Altman, 1986: 746), or to "mindless ... statistical rituals" (Gigerenzer, 2004). Nevertheless, you have to engage with significance testing because the corresponding p-values are abundant in the literature.[4]

Start by stating a null hypothesis (H_0) and a corresponding alternative hypothesis. When performing a significance test on the difference between two groups, this null hypothesis is usually:

H_0 : there is no difference between groups

The alternative hypothesis (H_A) is usually what the researcher actually believes in. In this case, the alternative hypothesis may be:

H_A : there is a difference between groups

In mathematical notation, this translates to:

$$H_0 : \mu_1 = \mu_2$$

$$H_A : \mu_1 \neq \mu_2$$

---

4 A disclaimer is needed here. It has to be emphasized that the procedure specified here is what has been adopted by the community; it does not follow the intentions of the original developers (see Perezgonzalez, 2015).

In other words, you *assume* that μ_1 and μ_2 are equal under the null hypothesis. Another way of writing down the null hypothesis is as follows:

$$H_0 : \mu_1 - \mu_2 = 0$$

In plain English: 'the difference between the two means is assumed to be zero' ... which is the same as saying that the two groups are *assumed* to be equal. Whether the null hypothesis is actually true or not is *not* the question. Notice one key detail in my notation: I am using Greek letters to specify the null hypothesis, which shows that the null hypothesis is an assumption about the population.

Once the null hypothesis has been stated for the population, you look at the particular sample from your study to measure how compatible or incompatible the actual data is with this initial assumption. That is, when engaging in NHST, you measure the incompatibility of the data with the null hypothesis. However, the only statements you are allowed to make are with respect to the data—whether the null hypothesis is actually true or not is out of your purview. The null hypothesis itself is an imaginary construct that you can never measure directly. I like to think about the null hypothesis as a 'statistical scapegoat' that you posit for the sake of argument. It's there to have something to argue against.

9.6. Using *t* to Measure the Incompatibility with the Null Hypothesis

Once the null and alternative hypothesis have been stated, you need to measure the incompatibility of the data with the null hypothesis. This is where 'test statistics' come into play. In 'cookbook' approaches to significance testing (see Chapter 16 for a critical discussion), there's a bewildering array of statistical tests, each with their own test statistic, such as t, F, χ^2, and more. Luckily—because this book doesn't endorse the testing framework—you don't have to learn all of these statistics. Learning about one of them, t, will suffice for our purposes. t is commonly used in the context of testing the difference between two groups.

$$t = \frac{\bar{x}_1 - \bar{x}_2}{SE} \tag{E9.5}$$

Notice how t combines all three of the ingredients mentioned earlier. The numerator contains the difference between two group means, which is the unstandardized effect size (the first ingredient). The denominator features the standard error, which, as was just discussed in the previous section, combines two ingredients—the variability in the data and the sample size. This formula should look remarkably similar to Cohen's d (see E9.1 above). The only difference is that t has the standard error rather than the standard deviation in the denominator. Because of this change, t cares about sample size (whereas Cohen's d doesn't). Figure 9.5 visualizes how the three ingredients impact t.

Very importantly, notice that I am using Roman rather than Greek letters in the formula for the t-value. This is because the t-value is an actual estimate that is generated

$$t = \frac{\bar{x}_1 - \bar{x}_2}{\dfrac{s}{\sqrt{N}}}$$

⟵ **t grows if the difference grows**

⟵ **t shrinks if the SD grows**

⟵ **t grows if N grows**

Figure 9.5. Three forces that impact t; more extreme t-values indicate a greater incompatibility with the null hypothesis of equal means[5]

from a sample. t will be our arbiter of significance—this will be the measure that is used to argue that the data is incompatible with the null hypothesis. However, we've now got a new problem: what's a small or a large t-value? And how big does a t-value have to be in order for us to conclude that the sample is incompatible with the null hypothesis? To be able to answer these questions, you need to know how t is distributed.

9.7. Using the *t*-Distribution to Compute *p*-Values

The statistic t actually has its own distribution under the null hypothesis, which is shown in Figure 9.6.[6] I won't go into the details of where this distribution comes from, but notice the striking similarity to the normal distribution. In fact, this distribution looks almost exactly like the normal distribution. The only difference is that the 'tails' (towards the left and the right) are a bit heavier compared to the normal, which means that very large or very small values are slightly more probable.

The t-distribution captures the probability of particular t-values if the null hypothesis of equal means were true. Given this, think about what the shape of the t-distribution encapsulates conceptually: The bell-shaped curve of this distribution essentially states that, under the null hypothesis of equal means, very large or very small t-values are very improbable. That is, only very rarely will random sampling result in samples with extreme t-values. In contrast, t-values closer to zero are much more probable under the null hypothesis of equal group means. This corresponds to the fact that in the absence of any group difference in the population (H_0), most randomly drawn samples are going to exhibit only very minor differences between two group means.

One way to think about this is to draw random samples from a distribution in which the null hypothesis is actually true. If you draw lots of samples, *most* selected samples will have t-values that are close to 0. Very rarely will random sampling produce extreme t-values.

The t-distribution can be used to compute the p-value. This number is a conditional probability, namely, the probability of the relevant test statistic (in this case, t) conditioned on the null hypothesis being true. For $t = 1.5$, the p-value is $p = 0.14$. This number is achieved by adding the area under the curve for the two shaded regions in

---

5 Technical side note: this is a simplified formula as the error in the denominator is computed differently for different types of t-tests (paired versus unpaired).
6 Technical side note: the t-distribution in Figure 9.6 is shown for 100 degrees of freedom.

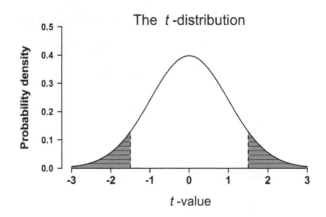

Figure 9.6. The *t*-distribution; striped areas indicate the probability of obtaining a *t*-value that is +1.5 or larger (right tail), and −1.5 or smaller (left tail); both areas taken together are equal to the *p*-value of a two-tailed significance test for an observed *t*-value of 1.5

Figure 9.6. That is, the *p*-value is the area under the curve for $t = 1.5$ or values more extreme than that, as well as for $t = -1.5$ and more extreme than that.[7]

Even though the number $p = 0.14$ seems pretty low, it is not enough to reject the null hypothesis for various historical reasons. The scientific community has converged on a rule where only *p*-values below the threshold of 0.05 are treated as good-enough evidence against the null hypothesis. This threshold is called the 'alpha level', with $\alpha = 0.05$ being the conventional threshold. If $p < \alpha$, the null hypothesis is rejected. The specific *t*-value that makes a *p*-value cross the α-threshold is also called a 'critical value'. The critical value turns out to be $t = 1.98$ in this case. If a sample generates a *t*-value that is as larger than this (or smaller than $t = -1.98$), *p* will be below 0.05.

Once $p < \alpha$, a result is claimed to be 'statistically significant', which is just the same as saying that the data is sufficiently incompatible with the null hypothesis. If the researcher obtained a significant result for a *t*-test, the researcher may act as if there actually was a group difference in the population.

Let us recap what you have learned about this procedure to this point. You define a population of interest, such as the population of all English speakers. You sample from this population of interest. When testing a difference between two groups, you state the null hypothesis $\mu_1 = \mu_2$ for the population. You compute a *t*-value from your sample. Finally, you investigate how improbable this *t*-value is under the null hypothesis of equal group differences. If $p < 0.05$, you act as if the null hypothesis is not true.

---

7 This is what's called a two-tailed test because it involves both tails of the *t*-distribution. In performing a two-tailed test, you disregard the sign of the *t*-value. That is, you are testing a non-directional hypothesis, which is standard procedure in most psychological and linguistic applications where people rarely consider one-tailed tests. As you want to be open to detecting significant effects that go the other way, performing two-tailed tests is generally preferred, and it is the more conservative option.

Notice that at no point in this procedure did you directly compute a statistic that relates to the alternative hypothesis. Everything is computed with respect to the null hypothesis. Researchers commonly pretend that the alternative hypothesis is true when $p < 0.05$. However, this is *literally* pretense because the significance testing procedure has only measured the incompatibility of the data with the null hypothesis.

9.8. Chapter Conclusions

This chapter started with the fundamental notion that in inferential statistics sample estimates are used to make inferences about population parameters. This chapter covered the basics of null hypothesis significance testing (NHST). A null hypothesis is posited that is assumed to characterize a phenomenon in the population, such as the means of two groups being equal ($\mu_1 = \mu_2$). Then, sample data is collected to see whether the sample is incompatible with this original assumption. Three ingredients influence one's confidence in rejecting the null hypothesis—the magnitude of an effect, the variability in the data, and the sample size. Standardized effect size measures such as Cohen's d and Pearson's r combine two of these (magnitude and variability), but they ignore sample size. Standard errors and confidence intervals combine variability and sample size. The test statistics used in significance testing (such as t) combine all three ingredients, and they are used to compute p-value. Once a p-value reaches a certain community standard (such as $p < 0.05$), a researcher may act as if the null hypothesis is to be rejected.

9.9. Exercises

9.9.1. Exercise 1: Gauging Intuitions About Cohen's d

In this exercise, you will generate some random data to gauge your intuitions about Cohen's d.

```
# Number of data points:
n <- 50

# Random y:

y <- c(rnorm(n, mean = 5, sd = 1),
    rnorm(n, mean = 2, sd = 1))

# Levels for x predictor:

x <- rep(c('A', 'B'), eac= n)

# Combine:

df <- tibble(x, y)

# Calculate Cohen's d:

cohen.d(y ~ x, df)
```

After implementing this code, play around with different means and different values for n. In addition, change the standard deviation of both groups, or one of the groups, to assess how this impacts *d*.

9.9.2. Exercise 2: Gauging Intuitions About *r*

In this exercise, you will generate some random data to gauge your intuitions about Pearson's *r*. The code is similar to Chapter 4.

```
x <- rnorm(50)

y <- 3 * x + rnorm(50, sd = 2)   # slope = 3

plot(x, y, main = cor(x, y))
```

Assess how changing the standard deviation and changing the slope impacts the correlation coefficient.

9.9.3. Exercise 3: Gauging Intuitions About *t* and Significance

In this exercise, you will generate some random data to gauge your intuitions about *t* and significance tests. The following performs a *t*-test for 10 participants. The group difference is specified to be 2, and the standard deviation is 2 as well. Notice that the following code uses the semicolon to stack up commands in the same line.

```
# Set values:

n <- 10; meandiff <- 2; my_sd <- 2

# Perform t-test:

t.test(rnorm(n, sd = my_sd),
       rnorm(n, sd = my_sd) + meandiff)
```

Run this entire batch of code multiple times. What *t*-values do you get? How often does the *t*-test become significant? Repeat this for different values for n, meandiff, and sd. This will show you how the three ingredients impact the test statistic and the resultant *p*.

10 Inferential Statistics 2: Issues in Significance Testing

10.1. Common Misinterpretations of p-Values

This chapter deals with issues in significance testing. First, it's important to talk about some common misunderstandings of p-values, which are notoriously easy to misinterpret.

One common misinterpretation is that the p-value represents the probability of the null hypothesis being true. This is *not* the case. Always remember that the null hypothesis is an assumption—its truth cannot be known.

Another common misinterpretation is that the p-value represents the strength of an effect. The last chapter discussed standardized measures of effect size such as Cohen's d. The statistics used in significance testing, such as t, do in fact incorporate effect size, but they also take the sample size into account. Measures such as t and the resulting p compress all the three ingredients of significance discussed in Chapter 9 into a single number. Because the p-value combines all ingredients, you cannot simply 'read off' the contribution of any one of the ingredients. In particular, a very small effect may still have a low p-value if the sample size is large. Thus, it is possible that a 'significant' result is very 'insignificant' in terms of effect size. This is why it's important to always mention some measure of effect size alongside the results of significance tests (Cumming, 2012, 2014).

A final misinterpretation worth mentioning is the idea that if $p < 0.05$, then one is justified to believe more strongly in one's hypothesis. This is not the case for two reasons. First, the p-value measures the incompatibility of the data with the null hypothesis H_0, it doesn't allow any direct conclusions about the alternative hypothesis H_A. The second reason is that in line with frequentist statistical philosophy, $p < 0.05$ says nothing concretely about the data at hand, instead, acting in line with this threshold ensures that you make correct decisions 95% of the time in the long run (see Dienes, 2008: 76). Philosophically, this is very dissatisfying. It thus doesn't come as a surprise that the logic of $p < 0.05$ is routinely flipped, with researchers interpreting this number as if it is measuring how much the data supports their desired theory (rather than how incompatible the data is with H_0).

10.2. Statistical Power and Type I, II, M, and S Errors

Several errors can happen when engaging in null hypothesis significance testing. Spuriously significant results are called Type I errors. A Type I error involves obtaining a significant effect even though the null hypothesis is actually a true characteristic of the population. A Type II error involves *failing* to obtain a significant effect even though the null hypothesis is false. Type I errors are also known as false positives; Type II errors are also known as false negatives.

Table 10.1 helps to clarify Type I and Type II errors. The rows show two states of the world: one in which the null hypothesis is true, and one in which it is false. In an actual analysis, the state of the world is unknown. The columns show two different scenarios: one in which the researcher obtains a sample that leads to $p < 0.05$ for the stated null hypothesis, and one in which the researcher obtains a sample that leads to $p < 0.05$.

Let's create a Type I error in R, using the t.test() function, which tests whether a group difference is significant (see Appendix A). The command below compares two sets of random numbers, rnorm(10) and rnorm(10). The default for the rnorm() function is $\mu = 0$. This means that the command below compares two sets of random numbers drawn from a normal distribution with the same mean (i.e., it is the same distribution). The set.seed() command is merely used to ensure that you and I get the same 'random' numbers (see Chapter 6.4).

```
set.seed(42) # set random number seed
t.test(rnorm(10), rnorm(10)) # output not shown
t.test(rnorm(10), rnorm(10)) # output not shown
t.test(rnorm(10), rnorm(10)) # output not shown
t.test(rnorm(10), rnorm(10)) # p < 0.05
```

```
        Welch Two Sample t-test

data: rnorm(10) and rnorm(10)
t = 2.3062, df = 17.808, p-value = 0.03335
alternative hypothesis: true difference in means is not
equal to 0
95 percent confidence interval:
0.06683172 1.44707262
sample estimates:
mean of x mean of y
0.5390768 -0.2178754
```

Table 10.1. Type I errors (false positive) and Type II errors (false negatives) in hypothesis testing

		Results from your sample	
		p < 0.05	*p > 0.05*
State of the world (the population)	**Nothing is there** (H_0 is actually true)	Type I error	Correct decision (don't claim result)
	Something is there (H_0 is actually false)	Correct decision (claim result)	Type II error

Running the test exactly four times yielded a significant test result on the fourth try. Random sampling created a spuriously significant result. We know that this is a Type I error because the distributions have been specified to have the same means ($\mu_1 = \mu_2$), so technically there *shouldn't* be any difference. If you accept the conventional significance threshold, you expect to obtain a Type I error 5% of the time (1 in 20). Now you realize that setting the alpha level to $\alpha = 0.05$ is the same as specifying how willing one is to obtain a Type I error. It's possible to set the alpha level to a lower level, such as $\alpha = 0.01$. In this case, Type I errors are expected to occur 1% of the time.

It's important to keep frequentist philosophy in mind when thinking about these procedures. For any given result, you can never rule out a Type I error. Chance sampling may always create apparent patterns that lead to spuriously significant results. However, you have control over how often you are willing to commit a Type I error. In other words, you have certainty about the significance testing procedure working out in the long run, even though you will always remain uncertain about any given dataset at hand.

Next, let's use R to simulate Type II errors. The following code initializes one group to have a mean of 1 and another group to have a mean of 0. Thus, this time around there *is* a difference in the population, and any result that is $p > 0.05$ would constitute a false negative. This is the case for the third run of the t.test() function below, for which the $p -$ value is 0.057.

```
set.seed(42)
t.test(rnorm(10, mean = 1), rnorm(10, mean = 0))
t.test(rnorm(10, mean = 1), rnorm(10, mean = 0))
t.test(rnorm(10, mean = 1), rnorm(10, mean = 0))
```

```
      Welch Two Sample t-test
data: rnorm(10, mean = 1) and rnorm(10, mean = 0)
t = 2.0448, df = 16.257, p-value = 0.05742
alternative hypothesis: true difference in means is not
equal to 0
95 percent confidence interval:
-0.03405031 1.95683178
sample estimates:
mean of x mean of y
0.97978465 0.01839391
```

Although tantalizingly close to 0.05, if you follow the procedure strictly, you have *failed to reject the null hypothesis* given your pre-specified alpha level. You cannot accept this result as significant; otherwise, you would give up the whole certainty that is afforded by specifying the alpha level in advance. Some people talk about "marginally significant" effects if they obtain p-values close to 0.05, but such talk is betraying the whole procedure.

The Type II error rate is represented by the letter β, the probability of missing a real effect. The complement of β is what is called 'statistical power', $1 - \beta$, sometimes represented by the letter π ('pi'). The power of a testing procedure describes its ability to detect a true effect. Many researchers aim for $\pi > 0.8$; that is, an 80% chance of obtaining a significant result in the presence of a real effect.

Just like significance, statistical power is affected by the three ingredients (effect size, variability, sample size). This means that you have three ways of increasing power: increase the effect size (such as by making your experimental manipulations more extreme), decrease the variability in your sample (for example, by making the sample more homogeneous), or increase your sample size by collecting more data. Arguably, sample size is the easiest thing to control. Thus, in practice, researchers most often aim to increase statistical power by conducting studies with larger sample sizes.

It is quite dangerous to place too much emphasis on low-powered studies, regardless of whether the results are significant or not. In particular, low-powered studies are more likely to suffer from what are called Type M and Type S errors (two terms first introduced by Gelman and Carlin, 2014). A Type M error is an error in estimating the *magnitude* of an effect, such as when your sample indicates a much bigger effect than is actually characteristic of the population. A Type S error is even worse; it represents a failure to capture the correct sign of an effect. For example, your sample may indicate taste words to be more positive than smell words, but in fact it's the other way around in the population. For a useful discussion of Type M and Type S errors in linguistics, see Kirby and Sonderegger (2018). Increasing statistical power not only lowers one's chances of committing a Type II error, but it also lowers the Type M and Type S error rate.

Considering statistical power is of pivotal importance for the interpretation of research studies. Many linguistic studies have very small sample sizes, which makes theoretical conclusions based on these samples quite flimsy. A case in point is a phonetic phenomenon known as 'incomplete neutralization'. I won't go into the details on this (highly controversial) topic here, but the main issue is that there were lots of studies that found results in support of incomplete neutralization, as well as quite a few studies that failed to find an effect. This led to an ugly battle of some researchers saying that incomplete neutralization exists, and others saying that it doesn't exist. Nicenboim, Roettger, and Vasishth (2018) show that accumulating the evidence across studies indicates a reasonably strong signal for the presence of incomplete neutralization.

This also points to another problem of small sample sizes: what researchers call 'null results' ($p > 0.05$) are basically uninterpretable if the power of a study is low. This is often paraphrased by the aphorism 'absence of evidence is not evidence of absence'. This means that you cannot easily claim that something *doesn't* exist if you failed to find a significant effect. This is the thinking error that characterized a lot of the studies that purported that the above-mentioned phenomenon of incomplete neutralization doesn't exist. They based this claim on exceedingly small samples, sometimes having samples consisting of only five speakers and a few words. Given the low statistical power of those studies, it comes as no surprise that they failed to find a significant effect.

Let me use this opportunity to say that it is safe to say most studies in linguistics are underpowered, which is a very dangerous situation to be in for any scientific field (Ionnadis, 2005). In some cases, having low power is unavoidable, such as when doing fieldwork on a moribund language of which there are only a handful of speakers left. However, in many other cases, there are no good reasons for low participant numbers and the researcher could easily have collected more data. The payoff of collecting more data is great, as it means that theoretical conclusions are based on sound evidence. Simply put, larger samples are more informative.

But what's a small or large sample? And how can you calculate power? As mentioned before, statistical power is influenced by the raw magnitude of an effect, the variability of the phenomenon, and the sample size. The problem is that the true numerical values for such quantities as the effect size cannot be known. So, in order to estimate power, you need to form reasonable expectations about effect size, preferably coming from past research. You can also calculate power for multiple effect sizes (from best case to worst case) to determine a reasonable sample size.

That said, this book unfortunately doesn't guide you through any actual power calculations. This is because ready-made formulas for calculating statistical power only exist for simple significance tests, not for many of the more complex models considered in this book, such as linear mixed effects models (Chapters 14 and 15). To circumvent this bottleneck, statistical power can be *simulated* by generating random datasets that exhibit the expected properties. Unfortunately, this requires more R programming than can be taught in this book. I recommend reading Kirby and Sonderegger (2018) and Brysbaert and Stevens (2018) for useful recommendations for power simulations with linguistic applications. Brysbaert and Stevens (2018) discuss the `simr` package (Green & MacLeod, 2016), which facilitates power simulations for mixed models.

10.3. Multiple Testing

Since all significance tests have a positive Type I error rate, the more tests a researcher conducts, the more likely they are going to stumble across a Type I error. This is known as the 'multiple testing' or 'multiple comparisons' problem. The 'family-wise error rate' is the probability of obtaining at least one Type I error for a given number of tests.[1] In the following formula, k represents the number of tests conducted at the specified alpha level (in this case $\alpha = 0.05$).

$$FWER = 1 - (1 - 0.05)^k \tag{E10.1}$$

The $(1 - 0.05)^k$ term represents the probability of *not* committing a Type I error for k number of tests. In the simple case of doing just one test, this term becomes: $(1 - 0.05)^1 = 0.95$. This states something you already know: if the null hypothesis is actually true, there is a high chance of *not* obtaining a significant result (that's a good thing). One minus this number yields the Type I error rate, $1 - 0.95 = 0.05$.

To get a feel for this formula, let's implement it in R and compute the probability of obtaining a Type error when performing a single test ($k = 1$):

```
1 - (1 - 0.05) ^ 1
```

```
[1] 0.05
```

[1] There also is the related concept of the 'false discovery rate', which is the expected proportion of Type I errors out of all tests with significant results. There are other rates as well, with different methods of correcting for them. The topic of 'multiple comparisons' is vast and due to space limitations this book can only scratch the surface.

The formula exhibits more interesting behavior once the number of significance tests (k) is larger. Let's see what happens if a researcher were to conduct 2 or 20 tests:

```
1 - (1 - 0.05) ^ 2
```

```
[1] 0.0975
```

```
1 - (1 - 0.05) ^ 20
```

```
[1] 0.6415141
```

Notice how the family-wise error rate increases quite rapidly. For only 2 tests, your chances of obtaining at least one Type I error are already 10%. When performing 20 significance tests, the chances rise all the way to 64%. For 100 significance tests, it's virtually impossible not to obtain at least one spuriously significant result.

Multiple testing problems have disastrous consequences for the interpretation of research papers. Austin, Mamdani, Juurlink, and Hux (2006) demonstrated the perils of performing lots of significance tests by testing a series of health-related variables against people's zodiac signs. Doing so many tests (for all zodiac signs and lots of health measures), they were bound to find at least some significant results. And, in fact, people born under Leo had a higher probability of gastrointestinal hemorrhage, and Sagittarians had a higher probability of humerus fracture. However, given how many associations were explored, these "findings" are likely spurious.

Demonstrations of multiple comparisons problem can also be quite entertaining: Bennett, Baird, Miller, and Wolford (2011) published a paper on the 'Neural correlates of interspecies perspective taking in the post-mortem Atlantic salmon' in the mock '*Journal of Serendipitous and Unexpected Results*'. This paper showed that when a dead salmon is placed in a brain scanner, it shows neural responses to videos of human social interactions in certain areas of its dead brain. Of course, this is just a statistical artifact. Testing brain activation is a noisy process that has a positive Type I error rate, which means that performing lots of significance tests on individual brain regions increases the family-wise error rate.

In both the zodiac study and the dead salmon study, the spurious associations disappeared once the researchers 'corrected' for performing multiple tests. There is a vast literature on multiple comparisons corrections, with many different approaches. The overarching logic of these correction procedures is that they make significance tests more conservative depending on how many tests a researcher conducts. The most straightforward and most widely used procedure is Bonferroni correction. The Bonferroni method asks you to reset your alpha level depending on k, the number of tests. The alpha rate is simply divided by the number of tests: $\frac{\alpha}{k}$. Let's say you conducted two tests. In that case, your new Bonferroni-corrected alpha level is $\frac{0.05}{2} = 0.025$. With this new alpha level, a p-value has to be below 0.025 to be treated as significant. A p-value of, say, $p = 0.03$ would not count as significant anymore.

When performing this correction, it is sometimes confusing to an audience to see really low p-values that are not treated as significant. So alternatively, you can uplift the p-value in correspondence with the Bonferroni-corrected α-level. The following command uses the p.adjust() function to adjust a *p*-value of *p* =0.03 for two tests, which results in a new *p*-value of *p* = 0.06 (not significant based on the original alpha level).

```
p.adjust(0.03, method = 'bonferroni', n = 2)
```

```
[1] 0.06
```

Multiple comparisons corrections such as the Bonferroni method are not without their enemies, and there is considerable controversy about when using them is appropriate. There is a big literature on whether one should correct or not, as well as a big literature on which precise correction method is the best (e.g., Rothman, 1990; Nakagawa, 2004). The easiest solution to the multiple comparisons conundrum is to limit the number of tests at the design phase of your study. If your study only features a small number of tests, each of which tests a different hypothesis that is theoretically motivated, the issue of increasing the study-wise Type I error rate is much diminished.

10.4. Stopping rules

The final topic to be discussed has a similar logical structure to the multiple comparisons problem. Imagine a researcher who wants to conduct a psycholinguistic study but didn't decide in advance how much data they were going to collect. After having collecting data from 30 participants, the researcher failed to find a significant effect. So the researcher decides to add data from 30 more participants. Adding more data seems like a rather innocuous change—after all, more statistical power is better, isn't it?

The issue here is that the decision to add more data was based on whether the first test was significant or not. It's generally better to have more data, but it's problematic to make decisions about one's sample size contingent on having obtained a significant result or not.

There's one key fact that you need to understand in order to recognize why flexibility in sample size is dangerous; namely, if the null hypothesis is true, *p*-values are uniformly distributed between 0 and 1. In other words, if the null hypothesis is true, any value between 0 and 1 is equally probable!

The behavior of a researcher who repeatedly performs significance tests for incrementally increasing sample sizes can easily be simulated (see Simmons et al., 2011). Figure 10.1 was constructed by performing a *t*-test repeatedly for 10 participants, 11 participants, 12 participants, and so on. The data was initialized without an actual difference in the population. The resulting series of *p*-values is a random walk. Figure 10.1 shows two such random walks. For the solid line, the researcher obtains a significant result after adding eight participants to the sample. If data collection is aborted at this stage, an erroneously significant result is published. The simulation shows that, in this particular case, the *p*-value *would have gone up again* after having crossed the threshold.

Dienes (2008: 68–69) talks about the need for scientists to have a dedicated 'stopping rule', which is a rule that determines when data collection is completed. In other words, the sample size should be decided in advance. In psychology in particular, there has been a renewed interest in this stopping rule problem, and many journals

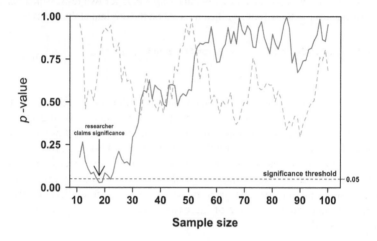

Figure 10.1. Assuming the null hypothesis is true, a *p*-value will eventually cross the significance threshold if the sample size is continuously increased; in the case of the solid line, the *p*-value *would have gone up again if more data was added*

now require authors to write a justification for the number of participants being tested. If you are pre-registering your study via a platform such as the Open Science Framework, you have to state the sample size in advance, as well as a rationale for why this sample size is appropriate. In some cases, it may be appropriate to argue for a given sample size based on previous research. For example, I just pre-registered a study with my student where we based our sample size specifications on the fact that we previously found significant effects in an experiment of the same sample size. The more principled way of arguing for a particular sample size is via a power analysis, for which Kirby and Sonderegger (2018) and Brysbaert and Stevens (2018) provide helpful pointers.

10.5. Chapter Conclusions

This chapter has made you aware of lots of issues in significance testing. It has dealt with Type I errors (false positives), Type II errors (false negatives), Type M errors (getting the magnitude of an effect wrong), and Type S errors (getting the sign of an effect wrong). In addition, the chapter introduced you to the issue of multiple comparisons, and the need to have a dedicated stopping rule that specifies when data collection is over. There are three take-home messages from this chapter. First, aim for high-powered studies. Second, don't conduct lots of theoretically unmotivated hypothesis tests without correcting your alpha level (your life will be easier if you limit the number of tests based on theoretical reasoning). Third, don't make data collection contingent on having obtained a significant result—decide in advance how many participants you want to run.

10.6. Exercise

10.6.1. *Exercise 1: Gauging Intuitions About the Bonferroni Method*

Imagine you obtained a *p*-value of $p = 0.001$ pretty low! But in fact, this *p*-value is just one of many. The actual study performed 100 significance tests. Using the Bonferroni method to correct the *p*-value for this number of tests yields $p = 0.1$, which is not significant anymore.

```
p.adjust(0.001, method = 'bonferroni', n = 100)
```

[1] 0.1

Play around with different *p*-values and different tests to get an intuition for how conservative the Bonferroni method is. You'll see that it's quite conservative! This is an important issue with multiple comparisons correction methods: the more tests you conduct, the more you have to lower your alpha level. As a result, statistical power shrinks as well.

11 Inferential Statistics 3
Significance Testing in a Regression Context

11.1. Introduction

The last two chapters have introduced you to null hypothesis significance testing. This chapter applies these concepts to regression modeling. This code-heavy chapter is more focused on implementation and less focused on concepts. You need to learn how to apply NHST to regression models and plot confidence intervals.

Remember that there always is uncertainty about where a particular parameter lies and whether your sample has estimated it correctly. There are two ways to communicate uncertainty about your models. One way involves communicating uncertainty about the parameter estimates (the regression coefficients, in particular). The second way is to communicate uncertainty about the predictions.

This chapter covers how to extract this information from linear model objects in R. Throughout the chapter, data from Chapters 2, 4 and 5 will be revisited.

11.2. Standard Errors and Confidence Intervals for Regression Coefficients

To start with, let's revisit the iconicity data from Winter et al. (2017). In Chapter 6, you created a model in which iconicity was conditioned on four different predictors: sensory experience ratings, imageability, systematicity, and log frequency. In addition, you standardized the different predictors to make the coefficients more comparable. The steps below repeat this analysis. There is one additional change compared to Chapter 6, which is applying the `format.pval()` function to the `p.value` column to make the output more digestible.[1] In addition, the code uses the `round()` function on the `estimate` and `std.error` columns to facilitate the following discussion.

```
library(tidyverse)
library(broom)
```

---

[1] When p-values focus on the SER column are very small numbers, R will display them in 'base-10 scientific notation'. For example, the number $8.678e-01$ translates into 0.8678. In comparison, $8.678e+02$ translates into 867.8. The 'e' stands for $\times 10^{exponent}$. So, 'e+02' means that you have to multiply by $10^2 = 100$; 'e-02' means that we have to *divide* by 100 (because $10^{-2} = 0.01$). Thus, this notation specifies by how much the decimal point has to be shifted around.

```
icon <- read_csv('perry_winter_2017_iconicity.csv')

icon %>% print(n = 4)


# A tibble: 3,001 x 8
  Word    POS          SER CorteseImag  Conc  Syst     Freq
  <chr>   <chr>      <dbl>       <dbl> <dbl> <dbl>    <int>
1 a       Grammatical NA            NA  1.46 NA     1041179
2 abide   Verb        NA            NA  1.68 NA         138
3 able    Adjective 1.73            NA  2.38 NA        8155
4 about   Grammatical 1.2           NA  1.77 NA      185206
# ... with 2,997 more rows, and 1 more variable:
#   Iconicity <dbl>
```

```
# Standardize predictors:

icon <- mutate(icon,
               SER_z = scale(SER),
               CorteseImag_z = scale(CorteseImag),
               Syst_z = scale(Syst),
               Freq_z = scale(Freq))

# Fit model:

icon_mdl_z <- lm(Iconicity ~ SER_z + CorteseImag_z +
                 Syst_z + Freq_z, data = icon)

# Look at coefficient table:

tidy(icon_mdl_z) %>%
  mutate(p.value = format.pval(p.value, 4),
         estimate = round(estimate, 2),
         std.error = round(std.error, 2),
         statistic = round(statistic, 2))
```

```
          term estimate std.error statistic p.value
1   (Intercept)     1.15      0.03     33.34 < 2e-16
2         SER_z     0.53      0.04     12.52 < 2e-16
3 CorteseImag_z    -0.42      0.04    -10.72 < 2e-16
4        Syst_z     0.06      0.03      1.79 0.07354
5        Freq_z    -0.36      0.11     -3.20 0.00142
```

Now that you know more about inferential statistics, we can unpack the full output of the regression table. First, there are the familiar `term` and `estimate` columns. Then there is the `std.error` column, which indicates how accurate you measure the corresponding coefficients. The `statistic` column contains *t*-values; the values are the `estimate` values divided by the respective standard errors. Finally, the column headed p.value displays *p*-values, which come from looking up the *t*-values on the *t*-distribution (see Chapter 9).

Importantly, whenever you see a coefficient table like this, you need to keep your eyes off the *p*-values until you have a firm grasp of the corresponding coefficients. If you don't understand the meaning of a particular coefficient, you may draw the wrong conclusions from the corresponding hypothesis test. This is particularly tricky when dealing with interactions (Chapter 8). Let's focus on the SER column for the time being to demonstrate the logic of these hypothesis tests. The *p*-value is very low (<2e-16), which translates into the following statement: 'Assuming that the SER slope is 0, the actually observed slope (+0.53) or any slope more extreme than that is very unexpected.'

What's most important for us right now is that you can use the standard errors to compute 95% confidence intervals for each regression coefficient. This is done by taking the estimate column and spanning a window of 1.96 times the standard error around this estimate (see Chapter 9). For the SER predictor, this interval is $[0.53 - 1.96*0.04, 0.53 + 1.96*0.04]$, which yields the 95% confidence interval [45,61] (with a little rounding).

When reporting the results of a regression model, it is conventional to report the coefficient estimate as well as the standard error. In my own papers, I often write statements such as the following: 'SER was positively associated with iconicity (+ 0.53, $SE = 0.04, p < 0.001$)'. I'd perhaps even describe the results in more detailed, conceptually oriented language: 'for each increase in sensory experience rating by one standard deviation, iconicity ratings increased by 0.53 ($b = 0.53, SE = 0.04, p < 0.001$)'.

Figure 11.1 shows a coefficient plot, otherwise known as 'dot-and-whisker plot', which plots each coefficient of the iconicity model with its corresponding 95% confidence interval. Essentially, this is a graphical display of a coefficient table. With its focus on confidence intervals, this plot encourages people to think about interval estimates, rather than just the point estimates for each regression slope (Cumming, 2012, 2014). This acts as a visual reminder that the point estimate is unlikely to capture the population parameter exactly.

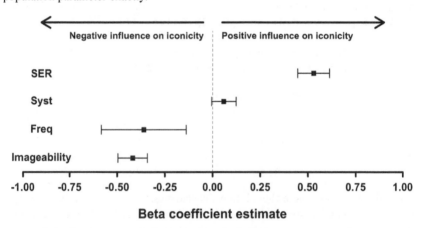

Figure 11.1. Point estimates of the regression coefficients (black squares) and their 95% confidence intervals

Let us create such a plot in R. First, handily, the `tidy()` function from the `broom` package can be used to create an output with 95% confidence intervals for each regression coefficient when specifying the argument `conf.int = TRUE`. In addition, the code below uses the `filter()` function to get rid of the row with the intercept, which we don't want to plot in this case.

```
mycoefs <- tidy(icon_mdl_z, conf.int = TRUE) %>%
  filter(term != '(Intercept)')
```

This tibble can then be used to create a dot-and-whisker plot, which will look similar to Figure 11.1, barring some differences to be explained below. This code may at first sight seem counterintuitive because of the use of `coord_flip()`. As the name suggest, this function flips the coordinates of a plot. The original *x*-axis becomes the new *y*-axis, and the original *y*-axis becomes the new *x*-axis. If this is confusing to you, run the code without the `coord_flip()` layer. This helps to understand why the `term` column is first mapped onto the *x*-axis, and why the estimates and the upper and lower bound of the confidence interval are mapped onto the *y*-axis, even though they ultimately end up on the *x*-axis thanks to `coord_flip()`.

```
mycoefs %>% ggplot(aes(x = term, y = estimate)) +
  geom_point() +
  geom_errorbar(aes(ymin = conf.low, ymax = conf.high),
                width = 0.2) +
  geom_hline(yintercept = 0, linetype = 2) +
  coord_flip() + theme_minimal()
```

This plot leaves a lot to be desired. Perhaps most importantly, the plot would look less 'random' if the coefficients were ordered in terms of their size, as is the case in Figure 11.1. This can be done by converting the `term` column into a factor, so that you can hand-specify a particular order of levels. The following code sorts the coefficient table in ascending order using the `arrange()` function and extracts the `term` column. The resulting `pred_order` vector contains the names of the slopes sorted by the size of the corresponding estimates. Then, the `factor()` function is used to transform the `term` column into a factor with the specified level order.

```
# Sort tibble by estimate and extract order of terms:

pred_order <- arrange(mycoefs, estimate)$term

pred_order
```

```
[1] "CorteseImag_z" "Freq_z"    "Syst_z"    "SER_z"
```

```
mycoefs <- mutate(mycoefs,
                  term = factor(term, levels = pred_order))
```

Now rerun the `ggplot()` command above and you will have a nicely sorted coefficient plot, as in Figure 11.1.

11.3. Significance Tests with Multilevel Categorical Predictors

What about significance testing if a categorical predictor has more than two levels? Let us revisit the modality data from Chapter 7. This analysis looked at context valence (whether a word occurred in overall good or bad contexts) as a function of taste and smell words. This section will expand this analysis to include all of the five senses.

Let's load the data back into your current R session.

```
senses <- read_csv('winter_2016_senses_valence.csv')

senses %>% print(n = 4)
```

```
# A tibble:   405 x 3
   Word       Modality    Val
   <chr>      <chr>       <dbl>
1 abrasive   Touch       5.40
2 absorbent  Sight       5.88
3 aching     Touch       5.23
4 acidic     Taste       5.54
# ... with 401 more rows
```

```
senses_mdl <- lm(Val ~ Modality, data = senses)
```

Let's look at the coefficient table. The following code uses rounding to facilitate discussion.

```
tidy(senses_mdl) %>%
  mutate(estimate = round(estimate, 2),
         std.error = round(std.error, 2),
         statistic = round(statistic, 2),
         p.value = format.pval(p.value, 4))
```

	term	estimate	std.error	statistic	p.value
1	(Intercept)	5.58	0.02	295.31	< 2.2e-16
2	ModalitySmell	-0.11	0.06	-1.93	0.05489
3	ModalitySound	-0.17	0.04	-4.64	4.663e-06
4	ModalityTaste	0.23	0.04	5.30	1.958e-07
5	ModalityTouch	-0.05	0.04	-1.21	0.22688

When interpreting the coefficient output, you have to remind yourself of the reference level, which is `Sight` in this case. The p-values in the right-hand column then correspond to the null hypothesis that the difference between `Sight` and `Smell` is equal to 0, that the difference between `Sight` and `Sound` is equal to 0, and so on.

Thus, the coefficient table only presents a very partial overview of the differences in the study.

To test the overall effect of sensory modality, you can perform a comparison of the model with `Modality` to a model without this predictor. In this particular case, the comparison model, which we can call the 'null model', is an intercept-only model because there are no other predictors. When doing this comparison, you are looking to see whether your model captures any variance in valence measures compared to the null model, what is sometimes called an 'omnibus test'.

To do this in R, let's first create the null model. Here it comes in handy to remember that '1' is a placeholder for the intercept.

```
senses_null <- lm(Val ~ 1, data = senses)
```

Next, the `anova()` function can be used for model comparison. The name of this function comes from 'analysis of variance' (ANOVA). This basic significance test assesses the variance that can be attributed to a factor of interest (such as `Modality`) against the overall variance. In the present case, this is equivalent to performing a model comparison of the model with the factor of interest against a model without the factor of interest. If the two models to be compared only differ in one predictor, then the difference in how much variance each model captures is due to that predictor.

```
# Perform model comparison:

anova(senses_null, senses_mdl)
Analysis of Variance Table
Model 1: Val ~ 1
Model 2: Val ~ Modality
  Res.Df    RSS Df Sum of Sq      F    Pr(>F)
1    404 33.089
2    400 28.274  4    4.8145 17.028 6.616e-13 ***
---
Signif. codes:
0 '***' 0.001 '**' 0.01 '*' 0.05 '.' 0.1 ' ' 1
```

You could report the results of this 'F-test' as follows: 'There was a statistically reliable effect of modality $(F(4,400) = 17.03, p < 0.0001)$'. Whereas t is a statistic used for group differences (t-tests) and regression coefficients, F is for comparing variances. Just like t, F has its own distribution under the null hypothesis that can be used to derive p-values. For the modality effect, the F-value is 17.03, which leads to a p-value much below 0.05. The numbers 4 and 400 are the 'degrees of freedom'. The number 400 corresponds to the total number of independent data points in this dataset.[2] The number 4 is the difference in the number of estimated parameters between

---

2 This number is always lower than the number of actual data points for reasons that I won't go into here. But, in general, the more parameters you estimate, the more degrees of freedom you lose.

the two models. The full model `senses_mdl` contains four more coefficients than the null model `senses_null`, which only contains an intercept.

In the case of having only one predictor, you can also use `glance()` to interrogate the overall model performance. The F-statistic and p-value in this output are the same as in the model comparison performed with `anova()`. This is not the case anymore when you have a model with multiple predictors. In that case, `glance()` performs a model comparison (F-test) of the entire model (including all predictors) against the model without any predictors (null model), which means that you can't associate p-values with any specific predictors anymore. You can paraphrase this comparison as asking the following question: assuming that the full model and the null model perform equally well (the null hypothesis), how surprising is the amount of sample variance explained by the full model? Or, more informally: how well do all predictors together capture variance in the response?

```
glance(senses_mdl)
```

```
   r.squared adj.r.squared    sigma statistic     p.value
1 0.1455037     0.1369588 0.2658678  17.02801 6.616243e-13
   df   logLik      AIC      BIC deviance df.residual
1   5 -35.62837 83.25674 107.2801 28.27428         400
```

So, you now know how to interpret the `glance()` output, which relates to the overall model performance. If you wanted to perform significance tests for specific multilevel predictors, you can use `anova()` for model comparison. In that case, you have to construct a model with and a model without the predictor in question. If you are not dealing with multilevel predictors, these extra steps won't be necessary. In the case of continuous and binary categorical predictors, you can simply rely on the significance tests that are reported in the coefficient output of the linear model.

In some fields, especially psychology, researchers are expected to assess the significance of all pairwise comparisons, in this case, sight versus sound, sight versus touch, sight versus taste, and so on. When doing this, you have to keep the multiple comparisons problem in mind (see Chapter 10): the more tests a researcher conducts, the more likely it is that any of these tests is significant. Let's perform a full sweep of pairwise comparisons for the senses data.

There are loads of packages that make it easy for you to compute pairwise comparisons. One particularly useful package for this is `emmeans` (Lenth, 2018). The following runs pairwise comparisons for all levels of the `Modality` predictor. The `adjust` argument specifies which particular adjustment method is used, in this case, Bonferroni correction.

```
library(emmeans)

emmeans(senses_mdl, list(pairwise ~ Modality),
        adjust = 'bonferroni')
```

```
$'emmeans of Modality'
 Modality   emmean         SE  df lower.CL upper.CL
 Sight    5.579663 0.01889440 400 5.542518 5.616808
```

```
Smell    5.471012 0.05317357 400 5.366477 5.575546
Sound    5.405193 0.03248092 400 5.341338 5.469047
Taste    5.808124 0.03878081 400 5.731884 5.884364
Touch    5.534435 0.03224121 400 5.471052 5.597818
```

```
Confidence level used: 0.95
```

```
$'pairwise differences of Modality'
contrast           estimate          SE  df t.ratio p.value
Sight - Smell   0.10865148 0.05643072 400   1.925  0.5489
Sight - Sound   0.17447036 0.03757671 400   4.643  <.0001
Sight - Taste  -0.22846083 0.04313872 400  -5.296  <.0001
Sight - Touch   0.04522812 0.03736969 400   1.210  1.0000
Smell - Sound   0.06581888 0.06230922 400   1.056  1.0000
Smell - Taste  -0.33711231 0.06581321 400  -5.122  <.0001
Smell - Touch  -0.06342336 0.06218459 400  -1.020  1.0000
Sound - Taste  -0.40293120 0.05058618 400  -7.965  <.0001
Sound - Touch  -0.12924225 0.04576577 400  -2.824  0.0498
Taste - Touch   0.27368895 0.05043260 400   5.427  <.0001
```

```
P value adjustment: bonferroni method for 10 tests
```

The function also spits out predictions, which will be covered later in this chapter. What concerns us now is the section that's headed $'pairwise differences of Modality'. After correction, there are six significant pairwise differences: sight/sound, sight/taste, smell/taste, sound/taste, sound/touch, and taste/touch.

However, I have to admit, I'm not a fan of the practice of performing tests for all comparisons within a study, even if these tests are Bonferroni-corrected. In particular, in the case of this dataset, I cannot think of any theoretical reasons for comparing the emotional valence of sight words and sounds words, or touch words and sight words. When there's no theoretical reason to perform a specific comparison, the analysis is essentially an exploratory one (see Chapter 16 for more on confirmatory versus exploratory analyses). In my paper on this data (Winter, 2016), I review literature which suggests that taste and smell words should be different from each other (taste words are overall more positive than smell), but the literature on this topic is not rich enough to allow making specific predictions for other pairwise contrasts. So, in the analysis reported in the paper, I opted to *only* perform a comparison between taste and smell words (exactly the analysis that you conducted in Chapter 7), and I did not correct the *p*-value of this result as it was just a single test.

There are other reasons for my dislike of performing a full suite of pairwise comparisons. For example, it is a philosophically thorny issue whether the significance of a specific binary contrast—say, A versus B—should depend on having performed a hypothesis test for a completely unrelated comparison, C versus D. In addition, there is the issue of statistical power: the more tests one conducts, the stricter Bonferroni correction becomes, which leads to a higher likelihood of committing a Type II error (a false negative). My final objection against performing all pairwise comparisons is that this practice buys into the common belief that all results should have *p*-values attached. There is already too much emphasis on these numbers, and performing pairwise comparisons furthers the practice of solely relying on *p*-values to make judgments about data.

Rather than performing a series of binary significance tests, I recommend interpreting your model with respect to the coefficients, the predictions, the effect sizes, etc. These are the things that are scientifically much more informative. In addition, as mentioned in Chapter 10, I recommend keeping the number of tests to be conducted low at the design phase of your study, formulating motivated hypotheses before you even collect the data, and then testing only these hypotheses.

11.4. Another Example: The Absolute Valence of Taste and Smell Words

As another example of my general approach to these sorts of problems, I will walk you through one more analysis from Winter (2016). Besides the prediction that taste words should be more positive than smell words, the literature on this topic affords making the prediction that taste and smell words together should be overall more emotionally engaging, or evaluative, than words for the other senses. To assess this, I used a measure I called 'absolute valence'. Remember that after z-scoring the data, quantities are expressed in terms of deviations from the mean. If one takes the absolute value (see Chapter 1.2) of this z-transformed measure, negative words are 'flipped over' and become positive. The resulting absolute valence measure then expresses the degree to which words assume extreme positions on the context valence scale—regardless of whether they are extreme in a 'good' or a 'bad' way. Words with low absolute valence scores are overall more neutral.

Let's reconstruct this measure and perform an omnibus test:

```
# Standardize valence and take the absolute value:

senses <- mutate(senses,
                 Val_z = scale(Val),
                 AbsVal = abs(Val_z))

# Omnibus test:

abs_mdl <- lm(AbsVal ~ Modality, data = senses)

# Model comparison without specifying null model directly:

anova(abs_mdl)
```

```
Analysis of Variance Table

Response: AbsVal
           Df  Sum Sq Mean Sq F value    Pr(>F)
Modality    4  14.611  3.6527  9.9715 1.061e-07 ***
Residuals 400 146.524  0.3663
---
Signif. codes:
0 '***' 0.001 '**' 0.01 '*' 0.05 '.' 0.1 ' ' 1
```

Notice that if the anova() function is wrapped around the model object, it automatically performs a test against the null model (glance() reports the same test). The

`Modality` factor is indicated to be significant for the new absolute valence measure. This is an important result that was reported in Winter (2016); however, the test doesn't tell us *which* of the five senses are significantly different from each other. Rather than performing all pairwise comparisons, I opted to only test the theoretically motivated hypothesis that taste and smell words *together* are different from sight, touch, and sound words. The following code achieves this by creating a new column, `ChemVsRest`, which separates the 'chemical senses' (taste and smell) from the rest. For this, the `ifelse()` function is used, which assigns the label `'Chem'` to all words that satisfy the logical statement 'Modality `%in%` chems', where the `chems` vector contains the labels for taste and smell words.

```
# Create taste/smell vs. sight/sound/touch predictor:

chems <- c('Taste', 'Smell')

senses <- mutate(senses,
                 ChemVsRest = ifelse(Modality %in% chems,
                                     'Chem', 'Other'))
```

Let's check that this has worked. The following command uses the `with()` function to make the `senses` tibble available to the `table()` function. This avoids having to re-type '`$senses`' multiple times (see Chapter 5.6).

```
with(senses, table(Modality, ChemVsRest))
```

```
         ChemVsRest
Modality Chem Other
   Sight    0   198
   Smell   25     0
   Sound    0    67
   Taste   47     0
   Touch    0    68
```

This works as a quick sanity check of the new `ChemVsRest` variable. Indeed, only taste and smell words are coded as 'Chem', and all other sensory modalities are coded as 'Other'. The new `ChemVsRest` variable can then be used as predictor in a linear model.

```
# Test this predictor:

abs_mdl <- lm(AbsVal ~ ChemVsRest, data = senses)

tidy(abs_mdl)
```

```
          term    estimate  std.error statistic
1    (Intercept) 1.0558262 0.07289898 14.483414
2 ChemVsRestOther -0.3422975 0.08039460 -4.257718
```

```
        p.value
1  1.440064e-38
2  2.572230e-05
```

As you can see, this comparison is significant. The slope is negative and, since 'Chem' is the reference level (it comes first in the alphabet), this means that words for the other senses were on average *less* valenced than words for the chemical senses.

In lumping sight, sound, and touch together as 'other', we have to recognize that this is quite a coarse comparison. However, it is a direct test of the idea that taste and smell together are less neutral than the other senses and, as such, it is a theoretically motivated comparison, in contrast to a full suite of pairwise comparisons.

11.5. Communicating Uncertainty for Categorical Predictors

Let's stay with this data for a little longer and use the predict() function to calculate predictions for each level of the Modality factor, as was already done in Chapter 7. For this specific case, we are using the senses_mdl (the model with the valence measure and all five senses, rather than the model with the absolute valence measure).

```
newpreds <- tibble(Modality =
                    sort(unique(senses$Modality)))

# Check:

newpreds
```

```
# A tibble: 5 x 1
  Modality
  <chr>
1 Sight
2 Smell
3 Sound
4 Taste
5 Touch
```

```
# Generate predictions:

fits <- predict(senses_mdl, newpreds)

fits
```

```
       1        2        3        4        5
5.579663 5.471012 5.405193 5.808124 5.534435
```

The object fits contains the predictions for the modalities in the newpreds tibble. To compute the 95% confidence interval by hand, the standard error for the predicted means is needed. For this, rerun the predict() function with the additional argument se.fit = TRUE. The resultant object is a named list. You can index this list with $se.fit to retrieve the standard errors.

```
# Standard errors for predictions:

SEs <- predict(senses_mdl, newpreds,
               se.fit = TRUE)$se.fit

SEs
```

```
         1          2          3          4          5
0.01889440 0.05317357 0.03248092 0.03878081 0.03224121
```

You now have two new vectors in your working environment: fits for the fitted values, and SEs for the corresponding standard errors. Let's put them both into the same tibble, which makes it easier to compute the 95% confidence intervals.

```
CI_tib <- tibble(fits, SEs)

CI_tib
```

```
# A tibble: 5 x 2
   fits    SEs
  <dbl>  <dbl>
1  5.58 0.0189
2  5.47 0.0532
3  5.41 0.0325
4  5.81 0.0388
5  5.53 0.0322
```

Now that everything is in the same tibble, mutate() can be used to compute the 95% confidence interval, which is roughly two times the standard error on both sides of the mean (see Chapter 9.4).

```
# Compute CIs:

CI_tib <- mutate(sense_preds,
                 LB = fits - 1.96 * SEs, # lower bound
                 UB = fits + 1.96 * SEs) # upper bound

CI_tib
```

```
# A tibble: 5 x 4
   fits    SEs    LB    UB
  <dbl>  <dbl> <dbl> <dbl>
1  5.58 0.0189  5.54  5.62
2  5.47 0.0532  5.37  5.58
3  5.41 0.0325  5.34  5.47
4  5.81 0.0388  5.73  5.88
5  5.53 0.0322  5.47  5.60
```

It's good to know how to perform these computations by hand; however, the `predict()` function can actually calculate the 95% confidence interval around the predicted means automatically in one go when the argument `interval = 'confidence'` is specified.[3] In the output below, `lwr` and `upr` are the lower and upper bounds of the 95% confidence interval around the means, respectively.

```
sense_preds <- predict(senses_mdl, newpreds,
                            interval = 'confidence')

sense_preds
```

```
        fit      lwr      upr
1 5.579663 5.542518 5.616808
2 5.471012 5.366477 5.575546
3 5.405193 5.341338 5.469047
4 5.808124 5.731884 5.884364
5 5.534435 5.471052 5.597818
```

The numbers are different due to rounding. However, you also need to be aware of the fact that depending on the sample size, `predict()` will compute the 95% confidence interval based on values that are slightly different from 1.96 (for reasons that won't be discussed here). In most cases, this will not make a big difference. If you want to be on the safe side, use the `predict()` function.

One more step is needed to create a plot of the means with their respective 95% confidence intervals. To plot the categorical modality labels on the *x*-axis and the predictions on the *y*-axis, the tibble needs to contain the modality labels in a separate column. For this, you can bind the `newpreds` tibble (which contains the modality labels) together with the `sense_preds` predictions, using the `cbind()` function. This function takes two two-dimensional R objects (such as two tibbles) and glues them together column-wise. A requirement for using this function is that the two objects have the same number of rows, which is in fact the case here.

```
sense_preds <- cbind(newpreds, sense_preds)

sense_preds
```

```
# A tibble: 5 x 4
  Modality    fit    lwr    upr
  <chr>     <dbl>  <dbl>  <dbl>
1 Sight      5.58   5.54   5.62
2 Smell      5.47   5.37   5.58
3 Sound      5.41   5.34   5.47
```

---

3   You can also specify the argument to be 'prediction', which will compute what is called a 'prediction interval'. This interval is different from a confidence interval. Whereas the confidence interval expresses uncertainty with respect to the *means*, the prediction interval expresses uncertainty with respect to future observations. That is, the 95% prediction interval tells you what values are plausible for the next data point sampled. Prediction intervals are always wider than confidence intervals.

```
4 Taste      5.81   5.73   5.88
5 Touch      5.53   5.47   5.60
```

Finally, everything is in place to create a plot of the predicted means and their 95% confidence intervals. The following command produces Figure 11.2 (left plot).

```
sense_preds %>%
  ggplot(aes(x = Modality, y = fit)) +
  geom_point() +
  geom_errorbar(aes(ymin = lwr, ymax = upr)) +
  theme_minimal()
```

Notice that this command defines a new set of aesthetic mappings for geom_errorbar(). These mappings assign the *y*-minimum of the error bar to the lower-bound column from the sense_preds tibble (LB), and the *y*-maximum of the error bar to the upper bound (UB).

The resulting plot leaves a number of things to be desired, both from an aesthetic as well as from a functional perspective. First, the modalities are shown in alphabetical order. It would be much nicer to show everything in order of increasing valence, from the most negative to the most positive modality. For this, you can recode the factor Modality to be in the order from the least to the most positive (ascending order). The following commands achieve this by extracting the Modality column from the sorted tibble.

```
# Extract ascending order:

sense_order <- arrange(sense_preds, fit)$Modality

# Set factor to this order:
```

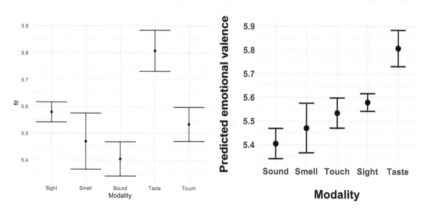

Figure 11.2. Predicted context valence and 95% confidence intervals; left: unordered, right: ordered and snazzy

```
sense_preds <- mutate(sense_preds,
                      Modality = factor(Modality,
                                        levels >= sense_order))
```

Next, there are a number of cosmetic improvements: the *x*- and *y*-axis labels could be increased in size to be more readable. The width of the error bars could be reduced to decrease the overlap between the different modalities. It also makes sense to increase the font size of the axis labels. Notice one little quirk in the command below, which is the '\n' in xlab() and ylab().The character sequence '\n' is interpreted as a line break. Including it in the axis text is a quick and dirty way of increasing the distance *x*-axis and *y*-axis labels. All of the code below produces the plot to the right of Figure 11.2.

```
sense_preds %>%
  ggplot(aes(x = Modality, y = fit)) +
    geom_point(size = 4) +
    geom_errorbar(aes(ymin = lwr, ymax = upr),
                  size = 1, width = 0.5) +
    ylab('Predicted emotional valence\n') +
    xlab('\nModality') +
    theme_minimal() +
    theme(axis.text.x =
            element_text(face = 'bold', size = 15),
          axis.text.y =
            element_text(face = 'bold', size = 15),
          axis.title =
            element_text(face = 'bold', size = 20))
```

11.6. Communicating Uncertainty for Continuous Predictors

What about plotting predictions for a continuous predictors? Let's revisit the English Lexicon Project data that was discussed in Chapters 4, 5, and 6. You will now recreate the plot of response times against frequency, with hand-specified 95% confidence intervals.

Let's redo the first steps of the analysis:

```
ELP <- read_csv('ELP_frequency.csv')

# Log-transform frequency predictor:

ELP <- mutate(ELP, Log10Freq = log10(Freq))

ELP
```

```
# A tibble: 12 x 4
   Word        Freq      RT  Log10Freq
   <chr>      <int> <dbl>      <dbl>
 1 thing      55522  622.       4.74
 2 life       40629  520.       4.61
 3 door       14895  507.       4.17
 4 angel       3992  637.       3.60
 5 beer        3850  587.       3.59
 6 disgrace     409  705        2.61
 7 kitten       241  611.       2.38
 8 bloke        238  794.       2.38
 9 mocha         66  725.       1.82
10 gnome         32  810.       1.51
11 nihilism       4  764.       0.602
12 puffball       4  878.       0.602
```

```
# Create linear model:

ELP_mdl <- lm(RT ~ Log10Freq, ELP)
```

Next, you need to define the data to generate predictions for. Let's generate a sequence of log frequencies from 0 to 5.

```
newdata <- tibble(Log10Freq = seq(0, 5, 0.01))
```

This can be used to compute predictions with predict(). As in the previous example, interval = 'confidence' ensures that 95% confidence interval around the predictions is computed.

```
preds <- predict(ELP_mdl, newdata,
                 interval = 'confidence')

head(preds)
```

```
       fit      lwr      upr
1 870.9054 780.8520 960.9588
2 870.2026 780.4127 959.9926
3 869.4999 779.9732 959.0266
4 868.7971 779.5334 958.0608
5 868.0943 779.0935 957.0952
6 867.3916 778.6534 956.1298
```

For plotting purposes, the fitted values should be stored in the same R object.

```
preds <- cbind(newdata, preds)

head(preds)
```

	Log10Freq	Log10Freq	fit	lwr	upr
1	0.00	0.00	870.9054	780.8520	960.9588
2	0.01	0.01	870.2026	780.4127	959.9926
3	0.02	0.02	869.4999	779.9732	959.0266
4	0.03	0.03	868.7971	779.5334	958.0608
5	0.04	0.04	868.0943	779.0935	957.0952
6	0.05	0.05	867.3916	778.6534	956.1298

Finally, everything is in place for plotting. The following plot (Figure 11.3) will be your first `ggplot2` that draws from more than one tibble—namely, the tibble of the actual data and the tibble of the predictions. Notice how, in the code below, `geom_text()` is specified to draw from a different tibble (ELP) than the other geoms, which draw from `preds`. The `geom_ribbon()` is used to plot the confidence region, with aesthetic mappings for the lower (`ymin`) and upper (`ymax`) boundary of the region. The `alpha` argument ensures that the confidence region is transparent. Finally, notice that the text is plotted *after* the ribbon layer so that the text is not occluded by the ribbon.

```
preds %>% ggplot(aes(x = Log10Freq, y = fit)) +
  geom_ribbon(aes(ymin = LB, ymax = UB),
              fill = 'grey', alpha = 0.5) +
  geom_line() +
  geom_text(data = ELP, aes(y = RT, label = Word)) +
  theme_minimal()
```

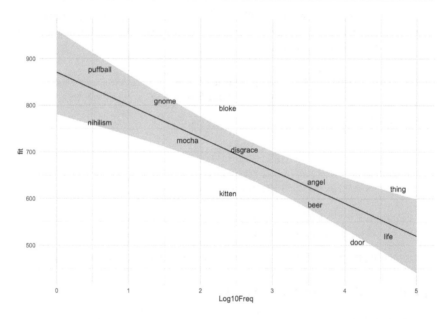

Figure 11.3. Scatterplot and regression fit, with the confidence region created with `geom_ribbon()`

How do you interpret the confidence region? As mentioned before, when estimating the population intercept and slope, there is always going to be variation between samples. Anytime that you draw a different sample of words (or of speakers), your coefficient estimates will differ, and so will the predictions based on these coefficients. I invite you to imagine lots of possible lines that you could have obtained, each one based on a slightly different intercept and a slightly different slope (sampling variation). We expect most of these lines to go through the gray region spanned by the 95% confidence interval that surrounds the regression line. This is the regression equivalent of the "dance of the confidence intervals" (Cumming, 2012, 2014) discussed in Chapter 9. Thus, you can use the confidence region as a visual reminder of the fact that there's uncertainty in the actual position of the regression line.

11.7. Chapter Conclusions

This chapter discussed the interpretation of the significance of regression coefficients. In addition, the chapter told you how to extract information about the uncertainty of regression coefficients and the uncertainty of predictions from model objects in R. You created a dot-and-whisker plot of regression slopes, with confidence intervals around the regression coefficients. In addition, you calculated the predictions with confidence intervals for various types of models in R. The skills discussed in this chapter give you greater flexibility in plotting your models. The best plots usually combine data and model.

11.8. Exercise

11.8.1. Exercise 1: Creating a (Bad) Coefficient Plot of Unstandardized Estimates

In this chapter, you created a dot-and-whisker plot for *standardized* regression coefficients of the iconicity model. What if you used the *unstandardized* regression coefficients instead? Recreate the model without the predictors labeled '_z', then plot the coefficients. What's bad about this plot? This should drive home some of the points made in Chapter 5 about why it's important to standardize in some contexts.

12 Generalized Linear Models 1
Logistic Regression

12.1. Motivating Generalized Linear Models

All of the models considered up to this point dealt with continuous *response* variables. Chapter 7 showed you how to incorporate categorical *predictors*. But what if the *response* itself is categorical?

Figure 12.1 is an example of such a categorical response. This is hypothetical data (but inspired by a real study: Schiel, Heinrich, & Barfüsser, 2012) in which the presence or absence of speech errors is modeled as a function of blood alcohol concentration (BAC): In this case, '0' corresponds to 'no speech error', and '1' corresponds to 'speech error'. The plot has a little *y*-scatter around the values 0 and 1 just to increase visibility of the individual points. The figure shows that on average, drunk people make more speech errors than sober people.

The curve in Figure 12.1 indicates the predicted probability of observing a speech error, based on what is called a 'logistic regression' model. This chapter will teach you how to fit such a model. Logistic regressions are ubiquitous in linguistics. For example, two-alternative forced choice responses or accuracy in psycholinguistics is often modeled using logistic regression. Other applications include: the presence or absence of a sociolinguistic variable (Drager & Hay, 2012; Tagliamonte & Baayen, 2012), the presence or absence of case marking (Bentz & Winter, 2013), or the choice between two types of syntactic constructions (Bresnan, Cueni, Nikitina, & Baayen 2007; Bresnan & Hay, 2008).

12.2. Theoretical Background: Data-Generating Processes

Before getting to logistic regression *per se*, I need to rewire your brain with respect to regression modeling. Remember that Chapter 4 claimed that in regression, the error is assumed to be normally distributed? It was also mentioned in Chapter 4 that this does *not* mean that the response variable itself has to be normally distributed. To clarify this, have a look at Figure 12.2. The thick solid line exhibits positive skew, but this distribution can actually arise from multiple normal distributions, shown by the dashed lines.

Fitting a model of the form 'y ~ Group' to this data will yield a prediction for each group. Crucially, the deviations from these predicted means, the residuals, will be normally distributed. This reinforces the idea that the normality assumption is not about the response variable *per se*, but about the residuals. There are situations in which the response looks skewed, but the normality assumption is satisfied.

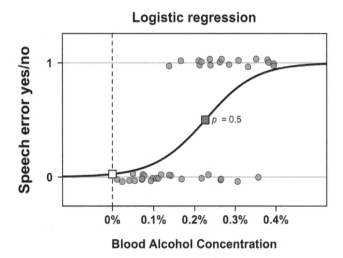

Figure 12.1. Speech errors as a function of blood alcohol concentration, treated as a binary categorical variable with superimposed logistic regression fit (bold curve); the white square indicates the intercept; the square in the middle indicates the point where making a speech error becomes more likely than not making a speech error

Now we get to rewiring your brain. Another way to think about this is that the normality assumption is not about the residuals *per se*, but about the process that has generated the data. Specifically, when fitting a linear model, you assume that the response variable *y* has been generated by a Gaussian (normal) process. Thus, the statement 'we assume Gaussian error' (residuals) and the statement 'we assume the data has been generated by a Gaussian process' go hand in hand. If values are drawn from an underlying normal distribution, the residuals will also be normally distributed.

Figure 12.2b encapsulates these ideas. First, let's unpack the $y \sim Normal\ (\mu, \sigma)$ part, ignoring the subindices (*i*) for the time being. This formula can be paraphrased as *y* is assumed to be generated by a normally distributed process with the mean μ and the standard deviation σ. In having been generated by a process that is normally distributed, the residuals are also going to be normally distributed around μ, with the spread given by σ.

The mean μ can then be conditioned on one or more predictors, such as *x*. The subindex *i* in Figure 12.2b is a placeholder, representing the fact that there are different values of *x*. For example, the first data point is $x_{i=1}$, the second is $x_{i=2}$, and so on. You can think of *i* as a counter variable, counting through the sequence of data points. For each data point, there is a different *i*. The subindex does something important in the formula, which is to make the *y*-values depend on the *x*-values. For example, x_1 may assume a different value from x_2, which then results in a different value for y_1 compared to y_2. In essence, this formula predicts a shifting mean; namely, a mean that shifts as a function of the *x*-values. *How* it shifts depends on the slope β_1. Notice that I am using Greek letters because everything said here relates to assumptions about the parameters. It is the job of regression to supply the corresponding estimates (b_1 for β_1, and so on).

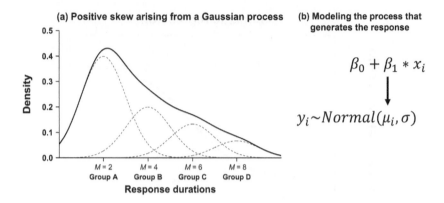

Figure 12.2. (a) Pooling multiple normal distributions (dashed lines) together may create the appearance of positive skew (thick solid line); (b) in linear regression, the data-generating process is assumed to be normally distributed

Essentially, everything is just as before—it's just that I invited you to think about *the process that has generated the data*. This way of thinking unlocks new potential. What if the process that has generated the data wasn't normally distributed? The 'generalized linear framework' *generalizes* the linear model framework to incorporate data-generating processes that follow any distribution. The first type of generalized linear model (GLM) that you will learn about is logistic regression, which assumes the response y to be binomially distributed, as shown in E12.1.

$$y \sim binomial(N = 1, p) \tag{E12.1}$$

The binomial distribution has *two* parameters, N and p. N is the 'trial' parameter; it describes how many trials are conducted. p is the probability parameter, and in this case, it describes the probability of y being either 0 and or 1. For our purposes, one parameter can be fixed, namely $N = 1$. In that case, the binomial distribution characterizes the probability of observing a single event, such as whether a speech error has occurred or not. In this chapter, you will use logistic regression exclusively to model data at the individual trial level, which is why you don't have to worry about the N parameter from now on. In fact, the binomial distribution with N set to 1 has a special name, it is called the 'Bernoulli distribution'. Thus, the formula E12.1 can be simplified to the following:

$$y \sim bernoulli(p) \tag{E12.2}$$

In other words, the way we are going to use it here, logistic regression assumes y to be generated by a process that follows a Bernoulli distribution. Figure 12.3a shows what probabilities the Bernoulli distribution assigns to the values '0' and '1' for three different parameters, $p = 0.2$, $p = 0.5$, and $p = 0.8$.

In the context of logistic regression, you are generally interested in modeling p as a function of one or more predictors. For example, you might want to model the

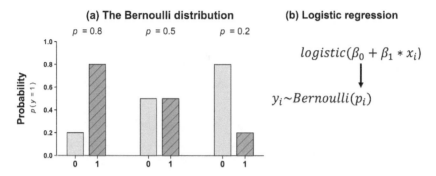

Figure 12.3. (a) The Bernoulli distribution for three different values of the parameter p, (b) logistic regression: assuming the response to be Bernoulli distributed and conditioning the parameter p on a set of predictors

probability of observing a speech error, $p(y = speech\,error)$, as a function of blood alcohol concentration. You might want to model the probability of passing a second language test, $p(y = pass)$, as a function of language background, age, and educational background. Or you might want to model the probability of a participant marking a word as perceptually prominent, $p(y = prominent)$, as a function of several acoustic variables, such as pitch and loudness (Baumann & Winter, 2018).

So, ultimately, you would want something like ' $p_i = \beta_0 + \beta_1 * x_i$', that is, you want different probabilities for different values of x. However, there's a snatch. The equation ' $\beta_0 + \beta_1 * x_i$' can predict *any* continuous value, but probabilities have to be between 0 and 1. Thus, you need a way of constraining what regression can predict; you need to 'crunch' the output of ' $\beta_0 + \beta_1 * x_i$' to fit into the interval [0,1]. For many areas of science (including statistics, computer science, artificial intelligence research), there's a certain 'go-to' function that is used when a continuous measure has to be compressed to the interval [0,1]. This function is the 'logistic function', and it lends logistic regression its name. So, rather than modeling the parameter p directly as a function of the predictors, the output of the predictive equation is transformed via the logistic, as shown in Figure 12.3b.

Figure 12.4 displays the effects of the logistic function. Notice how negative numbers such as –1 (bottom gray line) are positive and within the interval [0,1] after being transformed by the logistic function.

The following applies the logistic function to the example values –2, 0 and +2.

$$logistic(-2) \approx 0.12$$
$$logistic(0) = 0.5$$
$$logistic(+2) \approx 0.88$$

Notice that, just as Figure 12.4 suggests, the output of this function does not exceed the interval [0,1]. Notice furthermore that applying the logistic function to the numerical value 0 yields 0.5, which corresponds to the dashed line in Figure 12.4. In R, the logistic function is implemented in the command `plogis()`.

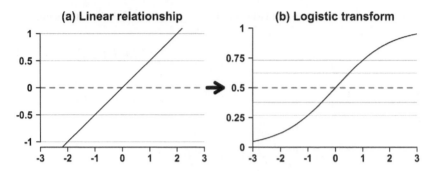

Figure 12.4. (a) A linear relationship between *y* and *x*; transforming the *y* s with the logistic function (b) restricts them to the interval between 0 and 1; the dashed line corresponds to a probability of 0.5 (graph inspired by McElreath, 2016)

```
plogis(-2)
```

```
[1] 0.1192029
```

```
plogis(0)
```

```
[1] 0.5
```

```
plogis(2)
```

```
[1] 0.8807971
```

I welcome you to plug in more numbers to get a feel for this function. Even if you supply very extreme values such as 10,000 or –10,000 to `plogis()`, it will always return a number between 0 and 1.

12.3. The Log Odds Function and Interpreting Logits

There's one more bit of math you need to learn about before you can start fitting your own logistic regression models. You need to know about 'log odds', otherwise called 'logits'. These are defined as follows:

$$\text{log odds} = \log\left(\frac{p}{1-p}\right) \tag{E12.3}$$

Here, *log* function is the natural logarithm (to the base *e*). The term inside the brackets are the odds, which are:

$$\text{odds} = \frac{p}{1-p} \tag{E12.4}$$

The odds express the probability of an event occurring (*p*) over the probability of an event *not* occurring (1 − *p*). You are actually familiar with the logic of odds from

Table 12.1. Representative probability, odds and log odds values; values rounded to two digits; notice that a probability of 0.5 corresponds to log odds of 0

Probability	Odds	Log odds ('logits')
0.1	0.11 to 1	−2.20
0.2	0.25 to 1	−1.39
0.3	0.43 to 1	−0.85
0.4	0.67 to 1	−0.41
0.5	**1 to 1**	**0.00**
0.6	1.5 to 1	+0.41
0.7	2.33 to 1	+0.85
0.8	4 to 1	+1.39
0.9	9 to 1	+2.20

everyday language. Almost certainly, you will have heard an expression such as 'the odds are one to one', which describes a 50% chance of an event occurring. You can express this statement by plugging $p = 0.5$ into equation E12.4, which yields:

$$\frac{0.5}{1-0.5} = \frac{0.5}{0.5} = 1 \tag{E12.5}$$

Why are the odds in E12.4 log-transformed? When you transform odds with the logarithm, you get a *continuous scale that ranges from negative infinity to positive infinity*. Table 12.1 shows the correspondence between probability, odds, and log odds for some representative values.

Log odds take considerable time to get used to—don't worry, there'll be lots of practice. A good thing to remember about log odds is that a log odds value of 0 corresponds to a probability of 0.5, and that positive log odds correspond to $p > 0.5$ and negative log odds correspond to $p < 0.5$. For example, if you are modeling the occurrence of speech errors, then a positive log odds value indicates that a speech error is more likely to occur than not.

The whole point of talking about log odds is that this puts probabilities onto a continuous scale, which is more amenable to being modeled with regression. Thus, logistic regression actually predicts log odds, as in E12.6. The shorthand 'logit' is used for 'log odds'.

$$logit(p_i) = \beta_0 + \beta_1 * x_i \tag{E12.6}$$

When wanting to report the model in terms of probabilities, you need to apply the logistic regression equation to the model's log odds predictions. The logistic function is the *inverse* of the log odd function.[1] Being each other's inverses, the log odds ('logit') function and the logistic function undo each other's effects, which is shown in Figure 12.5.[2]

1 Because the logistic is the inverse of the log odd, other courses use the notation $logit^{-1}$ for the logistic function, where the superscript −1 stands for 'inverse'.
2 You already know one pair of functions that are each other's inverse: the logarithm and the exponential function (Chapter 5). You can 'undo' the logarithmic transform by subsequently exponentiating. Likewise, you can 'undo' an exponentiation by subsequently applying the logarithm.

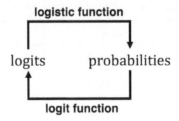

logistic function

logits   probabilities

logit function

Figure 12.5. The correspondence between probabilities and logits; the logistic function crunches logits into the range [0,1]; the logit function expresses probabilities on a scale that ranges from negative infinity to positive infinity

If you are new to logistic regression and not so mathematically inclined, the last section may have been hard. That's OK. The following presents three applications of logistic regression, and in walking you through the same procedures again and again, the correspondence between probabilities and log odds will eventually sink in.

12.4. Speech Errors and Blood Alcohol Concentration

Let's use the (artificial) blood alcohol concentration data shown in Figure 12.1 to fit your first logistic regression model. The data is contained in the file 'speech_errors.csv'.

```
library(tidyverse)
library(broom)

alcohol <- read_csv('speech_errors.csv')

alcohol
```

```
# A tibble: 40 x 2
      BAC speech_error
    <dbl>        <int>
 1 0.0737           0
 2 0.0973           0
 3 0.234            0
 4 0.138            1
 5 0.0933           0
 6 0.262            1
 7 0.357            0
 8 0.237            1
 9 0.352            1
10 0.379            1
# ... with 30 more rows
```

The column BAC contains the blood alcohol concentration predictor. The column speech_error contains information about the presence (1) or absence (0) of a speech error. This will be the response variable of your logistic regression model. The function for fitting a logistic regression model is called glm() (for generalized linear model). You specify your model formula as before, but in addition, you need to specify the assumed distribution of the data-generation process. This is done via the `family` argument. The name of this argument, 'family', comes from the fact that you can think of any basic distributional shape (e.g., uniform, Gaussian, binomial) as a family of distributions. This is because changing the parameters allows you to create lots of versions of the same distribution. You specify the family to be 'binomial' (remember that the Bernoulli distribution is a specific case of the binomial distribution).[3]

```
alcohol_mdl <- glm(speech_error ~ BAC,
                   data = alcohol, family = 'binomial')
```

Now that you have the fitted model stored in the object `alcohol_mdl`, you can use the `broom` function `tidy()` to retrieve the coefficient table—just as you're used to from linear regression.

```
tidy(alcohol_mdl)
```

```
        term   estimate std.error  statistic      p.value
1 (Intercept) -3.643444  1.123176 -3.243878 0.0011791444
2         BAC 16.118147  4.856267  3.319041 0.000903273
```

As always, most of our time should be devoted to interpreting the `estimate` column. In this case, the estimates are given in log odds. The first thing to look for is the sign of each coefficient. Notice that the slope of `BAC` is positive, which means that an increase in blood alcohol concentration corresponds to an increase in the log odds of observing a speech error. Notice furthermore that the sign of the intercept is negative, which indicates that for $x = 0$, it is the case that $p(y = speech\ error) < 0.5$. In other words, sober people make a speech error less than 50% of the time. The intercept is represented by the white square in Figure 12.1.

The fact that the p-value for the BAC coefficient is significant can be translated into the following statement: 'Assuming that the slope $BAC = 0$, obtaining a slope of 16.11 or more extreme than that is quite unlikely.' You could report this result as follows: 'There was a reliable effect of BAC (logit coefficient: +16.11, $SE = 4.86$, $z=3.3$, $p = 0.0009$).' Notice that the test statistic in the case of logistic regression coefficient turns out to be z rather than t (for reasons that I won't go in here).

To get rid of the confusing log odds, let's calculate some probabilities. Let's first extract the coefficients.

```
intercept <- tidy(alcohol_mdl)$estimate[1]

slope <- tidy(alcohol_mdl)$estimate[2]

intercept
```

```
[1] -3.643444
```

---

3 When using `glm()`, the arguments `family = 'binomial'`, `family = binomial()`, and `family = binomial(link = 'logit')` are equivalent.

```
slope
```

```
[1] 16.11815
```

Let's compute the log odds values for a blood alcohol concentration of 0% (completely sober) and for a blood alcohol concentration of 0.3% (*really* drunk).

```
intercept + slope * 0 # BAC = 0
```

```
[1] -3.643444
```

```
intercept + slope * 0.3 # BAC = 0.3
```

```
[1] 1.192
```

These are the predicted log odds for the corresponding blood alcohol concentrations. To get the predicted probabilities of making a speech error, apply the logistic function `plogis()` to these log odds.

```
plogis(intercept + slope * 0)
```

```
[1] 0.02549508
```

```
plogis(intercept + slope * 0.3)
```

```
[1] 0.7670986
```

For sober people, the predicted probability of a speech error is 0.025. Thus, given this model, you expect speech errors to occur on average about 2.5% of the time. For drunk people, the predicted probability is 0.77. Thus, you expect speech errors to occur on average about 77% of the time.

To recreate Figure 12.1, use the familiar `seq()` function to generate a series of *x*-values. Then use these *x*-values to generate predicted probabilities.

```
BAC_vals <- seq(0, 0.4, 0.01)

y_preds <- plogis(intercept + slope * BAC_vals)
```

Let's put both vectors into a tibble.

```
mdl_preds <- tibble(BAC_vals, y_preds)

mdl_preds
```

```
# A tibble: 41 x 2
   BAC_vals y_preds
      <dbl>   <dbl>
 1    0      0.0255
 2    0.01   0.0298
 3    0.02   0.0349
 4    0.03   0.0407
 5    0.04   0.0475
 6    0.05   0.0553
 7    0.06   0.0644
 8    0.07   0.0748
 9    0.08   0.0867
10    0.09   0.100
# ... with 31 more rows
```

Notice how the values in the `y_preds` column increase as the BAC values increase (higher blood alcohol concentrations correspond to more speech errors).

The following code reproduces Figure 12.1 with `ggplot()`. Notice that `geom_point()` draws from the mappings specified inside the `ggplot()` function, the `alcohol` tibble. `geom_line()`, on the other hand, draws its predicted values from the `mdl_preds` tibble (compare Chapter 11.6).

```
ggplot(alcohol, aes(x = BAC, y = speech_error)) +
  geom_point(size = 4, alpha = 0.6) +
  geom_line(data = mdl_preds,
            aes(x = BAC_vals, y = y_preds)) +
  theme_minimal()
```

12.5. Predicting the Dative Alternation

Let's perform another logistic regression, following the steps of Bresnan and colleagues (Bresnan et al., 2007). Linguists are interested in what is called the 'dative alternation'. For example, English speakers can either say, *Who gave you that wonderful watch?* or *Who gave that wonderful watch to you?* (Bresnan & Hay, 2008). The first syntactic construction is called a 'double object construction'. The second syntactic construction is called a 'prepositional dative'. What makes speakers choose one construction over another?

There are many predictors of the dative alternation. Here, only the role of 'animacy' will be explored. Compare the question *Who sent the box to Germany?* to the question *Who sent Germany the box?* The recipient of the sending event is Germany, an inanimate referent, in contrast to animate referents such as *Sarah, Bill,* or *the children.* For many speakers, *Who sent Germany the box?* sounds a bit strange. However, when the recipient is animate, the same double object construction seems to work just fine, as in *Who sent Sarah the box?* Bresnan and colleagues (2007) found that the prepositional dative was greatly preferred when the recipient was inanimate (*sent the box to Germany*).

The relevant dataset is called `dative`, and it is accessible via the `languageR` package (Baayen, 2013).

```
library(languageR)
```

Let's check the first two rows of the `dative` data frame with `head()` (this is not a tibble).

```
head(dative, 2)
```

```
  Speaker Modality Verb SemanticClass LengthOfRecipient
1   <NA>  written feed             t                 1
2   <NA>  written give             a                 2
  AnimacyOfRec DefinOfRec    PronomOfRec LengthOfTheme
1      animate   definite    pronominal            14
2      animate   definite nonpronominal             3
  AnimacyOfTheme DefinOfTheme PronomOfTheme
1       inanimate   indefinite nonpronominal
2       inanimate   indefinite nonpronominal
  RealizationOfRecipient AccessOfRec AccessOfTheme
1                     NP       given           new
2                     NP       given           new
```

The relevant response variable is `RealizationOfRecipient`. Let's check the content of this column with the `table()` function.

```
table(dative$RealizationOfRecipient)
```

```
  NP   PP
2414  849
```

There were 2,414 instances of the 'NP' construction (double object construction, *Who gave you that wonderful watch?*) and 849 instances of the 'PP' construction (prepositional dative, *Who gave that wonderful watch to you?*). We want to model this binary category as a function of `AnimacyOfRec`, which describes the animacy of the recipient (inanimate: *Germany*, versus animate: *Sarah*). The corresponding `glm()` function call looks as follows:

```
dative_mdl <- glm(RealizationOfRecipient ~ AnimacyOfRec,
                  data = dative, family = 'binomial')

tidy(dative_mdl)
```

```
                     term   estimate   std.error
1             (Intercept)  -1.154058  0.04259436
2 AnimacyOfRecinanimate    1.229407  0.13628810
     statistic       p.value
1    -27.09415  1.154003e-161
2      9.02065  1.869763e-19
```

Notice that because 'animate' comes before 'inanimate' in the alphabet, the animate recipient category is assigned to be the reference level of the animacy predictor. Thus, the predicted log odds for the animate recipients are 'hidden' in the intercept. You can also see this by the fact that the slope mentions 'inanimate' (AnimacyOfRecinanimate), which corresponds to the fact that this is a slope *towards* the animate category. Thus, the model predicts a log odds value of -1.15 for animates and a log odds value of $-1.15 + 1.23 = 0.08$ for inanimates. But wait, is this output given in terms of predicting the probability of prepositional datives or in terms of predicting the double object construction?

To answer this question, you need to know the order of the levels in the column of the response variable, for which you can use the `levels()` function (see Chapters 1 and 7).

```
levels(dative$RealizationOfRecipient)
```

```
[1] "NP" "PP"
```

Logistic regression in R will always model the quantity that is shown to the right, so, in this case, `"PP"`, the prepositional dative. Thus, the positive slope of $+1.22$ means that for inanimates (as opposed to animates), the odds of observing a prepositional dative increase. Let's calculate the probabilities that correspond to these log odds.

```
intercept <- tidy(dative_mdl)$estimate[1]

slope <- tidy(dative_mdl)$estimate[2]

plogis(intercept + slope * 0)
```

```
[1] 0.2397487
```

```
plogis(intercept + slope * 1)
```

```
[1] 0.5188285
```

Thus, the model predicts that the probability of observing a prepositional dative is 0.52 when the recipient is inanimate. It only predicts a probability of 0.24 when the recipient is animate.

How could you report this result? Here's one way of doing it: 'The predicted probability of observing a prepositional dative was 0.24 for animate recipients and 0.52 for inanimate recipients (logit difference: $+1.22$, $SE = 0.13$, $z = 9.02$, p< 0.0001).' To interpret the *p*-value for the AnimacyOfRecinanimate difference, you need to remember that 'e-19' means that the reported number (1.9) has to be shifted 19 decimal places to the right, which can be reported as p< 0.0001 (it is conventional to use such 'smaller than' statements for very low *p*-values). Thus, under the null hypothesis that there is no difference between inanimate and animate recipients, this data is fairly unexpected.

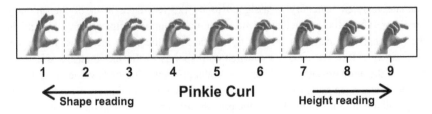

Figure 12.6. A nine-stepped continuum of 3D hand shapes that we used as stimuli for a web-based perception experiment in Hassemer and Winter (2016); the hand shapes were designed by 3D graphic designer Philip Krakow

12.6. Analyzing Gesture Perception

12.6.1. Exploring the Dataset

In the final example, you will analyze data from a gesture perception experiment conducted by Hassemer and Winter (2016) (see also Hassemer & Winter, 2018). In this study, we were trying to understand how onlookers infer information from gesture, using 3D hand shapes such as those shown in Figure 12.6. For the hand shapes towards the right in this figure, the pinkie finger, the ring finger, and the middle finger are curled in. For the hand shapes towards the left, these fingers are extended. In our paper, we called this variable 'pinkie curl'. In his PhD thesis, Hassemer (2016) claimed that hand gestures with a high degree of pinkie curl (with the pinkie curled in, to the right of Figure 12.6) lead to a 'height interpretation' of this gesture. In this case, the onlooker focuses on the distance between the index finger and the thumb pad, as if the gesturer was indicating the size of an imaginary object held between the fingers. In contrast, for the hand shapes towards the left of Figure 12.6, Hassemer (2016) argues that a shape reading is more likely, with the C-shape spanned between the index finger and thumb indicating a round shape.

In our experiment, we showed participants a single hand shape from the continuum in Figure 12.6. We then asked participants whether they thought the gesture indicated the height or the shape of an object. This is a two-alternative forced choice task, with the choice between 'height' and 'shape' being our response measure. The goal of the following analysis is to model this response as a function of 'pinkie curl'. Let's load the data into your current R session.

```
ges <- read_csv('hassemer_winter_2016_gesture.csv')

ges
```

```
# A tibble: 309 x 5
   index_curve pinkie_curl question_order confidence choice
         <int>       <int> <chr>               <int> <chr>
1            1           9 height_first            8 height
2            5           6 shape_first             7 shape
3            2           7 height_first            8 height
4            4           3 height_first            8 shape
```

5	1	1 height_first	8	shape
6	6	9 height_first	9	shape
7	6	7 shape_first	6	shape
8	3	7 shape_first	7	shape
9	1	2 shape_first	3	shape
10	6	2 height_first	7	shape

```
# ... with 299 more rows
```

Each row in this tibble represents one participant (there were 309 participants in total). For now, the only relevant columns are `pinkie_curl` (the predictor) and `choice` (the response). As usual, start by familiarizing yourself with this dataset. How were the participants distributed across the pinkie curl conditions? To answer this question, you can tabulate the counts of data points per pinkie curl value with the `table()` function.

```
table(ges$pinkie_curl)
```

```
 1  2  3  4  5  6  7  8  9
33 37 37 30 42 44 32 24 30
```

As shown in Figure 12.8, there are 9 steps to the pinkie curl continuum, ranging from 1 (fingers extended) to 9 (fingers curled in). Based on this table, it looks like there are about equally many participants per condition. Next, which response option was chosen more frequently?

```
table(ges$choice)
```

```
height shape
   125   184
```

These counts suggest that participants overall preferred the shape option. You can check the proportion by dividing the two cell counts by the total count:

```
table(ges$choice) / sum(table(ges$choice))
```

```
   height     shape
0.4045307 0.5954693
```

For the same result, you can put `prop.table()` around the `table()` output.

```
prop.table(table(ges$choice))
```

```
   height     shape
0.4045307 0.5954693
```

Let us cross-tabulate the two response options against the pinkie curl variable and store the resulting contingency table in an object called `xtab`.

```
xtab <- table(ges$pinkie_curl, ges$choice)

xtab
```

```
  height shape
1     14    19
2     17    20
3     12    25
4      5    25
5      9    33
6     15    29
7     15    17
8     18     6
9     20    10
```

Comparing the counts in the left column (height responses) to the counts in the right column (shape responses) for different pinkie curl values reveals that there are more height than shape responses *only* for the two biggest pinkie curl values (8 and 9). In other words, shape responses are relatively underrepresented for high pinkie curl variables, just as Hassemer's (2016) theory predicts. This pattern is perhaps easier to see when looking at row-wise proportions. The rowSums() function can be used to divide each count by the sum of each row.

```
xtab / rowSums(xtab)
```

```
      height       shape
1  0.4242424   0.5757576
2  0.4594595   0.5405405
3  0.3243243   0.6756757
4  0.1666667   0.8333333
5  0.2142857   0.7857143
6  0.3409091   0.6590909
7  0.4687500   0.5312500
8  0.7500000   0.2500000
9  0.6666667   0.3333333
```

Alternatively, you can use prop.table() and specify '1' to compute row-wise proportions. The following code also uses round() to round the proportions to two digits.

```
round(prop.table(xtab, 1), digits = 2)
```

```
  height shape
1   0.42  0.58
2   0.46  0.54
3   0.32  0.68
4   0.17  0.83
5   0.21  0.79
```

```
6    0.34    0.66
7    0.47    0.53
8    0.75    0.25
9    0.67    0.33
```

12.6.2. Logistic Regression Analysis

Now that you have a good understanding of this dataset, let's fit a logistic regression model. You want to model `choice` as a function of `pinkie_curl`. Type in the following line of code, which results in an error message.

```
ges_mdl <- glm(choice ~ pinkie_curl, data = ges) # error
```

```
Error in y - mu : non-numeric argument to binary operator
```

The cryptic error message tells you that something is wrong with 'y'. The problem is that you forgot to specify the `family` argument, which needs to be set to `'binomial'` for logistic regression—otherwise, the `glm()` function does not know what type of generalized linear model to fit.

```
ges_mdl <- glm(choice ~ pinkie_curl,
                data = ges, family = 'binomial') # error
```

```
Error in eval(expr, envir, enclos) : y values must be 0 <= y <= 1
```

Another error message! The problem here is that `choice` is coded as a character vector. However, for logistic regression, the response needs to be either coded as 0 or 1, or it needs to be coded as a factor.

```
# Convert to factor and check:

ges <- mutate(ges, choice = factor(choice))

class(ges$choice)
```

```
[1] "factor"
```

It's worth checking in which order the levels are listed:

```
levels(ges$choice)
```

```
[1] "height" "shape"
```

This means that logistic regression will report the probability of observing a 'shape' response. So, let's fit the model:

```
ges_mdl <- glm(choice ~ pinkie_curl, data = ges,
                family = 'binomial')

tidy(ges_mdl)
```

```
            term      estimate  std.error statistic       p.value
1 (Intercept)      1.0651620 0.26714076  3.987269 6.683834e-05
2 pinkie_curl     -0.1377244 0.04794959 -2.872274 4.075298e-03
```

Notice that `pinkie_curl` is entered as a numeric variable.[4] Because `pinkie_curl` is a numeric predictor, you can interpret the corresponding slope to mean that, for each increase in `pinkie_curl` by one step along the pinkie curl continuum, the log odds of observing a shape response decrease by -0.13772. To put this into plain English: as the pinkie finger becomes more curled in, shape responses become less likely. For interpretative purposes, you can also reverse the sign—in which case, the log odds of observing a *height* response increase by $+0.13772$ for each step along the continuum.

To generate predictions for all pinkie curl values, `predict()` comes in handy. Let's generate a tibble with values to generate predictions for.

```
ges_preds <- tibble(pinkie_curl = 1:9)

predict(ges_mdl, ges_preds)
```

```
        1          2           3           4          5
0.9274376  0.7897133   0.6519889   0.5142645  0.3765402
        6          7           8           9
0.2388158  0.1010915  -0.0366329  -0.1743573
```

By default, `predict()` returns log odds. Notice that the predicted log odds are only negative for the very high pinkie curl values of 8 and 9, which corresponds to the fact that shape responses are *less* likely than height responses only for very high pinkie curl values. To compute probabilities, use the logistic function `plogis()`.

```
plogis(predict(ges_mdl, ges_preds))
```

```
        1          2          3          4          5
0.7165551  0.6877698  0.6574585  0.6258056  0.5930384
        6          7          8          9
0.5594218  0.5252514  0.4908428  0.4565208
```

Notice how the values for 8 and 9 are below 0.5, in line with the fact that the corresponding log odds were negative.

An alternative way to compute these probabilities is to use the `type = 'response'` argument from the `predict()` function. This saves you using `plogis()`.

```
predict(ges_mdl, ges_preds, type = 'response')
```

```
        1          2          3          4          5
0.7165551  0.6877698  0.6574585  0.6258056  0.5930384
        6          7          8          9
0.5594218  0.5252514  0.4908428  0.4565208
```

---

4 This is legitimate here because the step-size between the different steps along the pinkie curl continuum is constant; that is, 1 and 2 are as far away from each other as are 8 and 9.

Let's compute the 95% confidence interval for plotting. The following code creates a tibble with the lower bound (`LB`) and upper bound (`UB`) of the confidence interval (see Chapter 11). Notice that, to compute the confidence interval in terms of probabilities, you first need to compute the full confidence interval in log odds. Once you have the log odd confidence interval, you can back-transform the lower bound and upper bound of log odds into probabilities. Don't transform the standard error and the fitted values separately.

```
ges_preds <- as_tibble(predict(ges_mdl,
                               ges_preds,
                               se.fit = TRUE)[1:2]) %>%
  mutate(prob = plogis(fit),
         LB = plogis(fit - 1.96 * se.fit),
         UB = plogis(fit + 1.96 * se.fit)) %>%
  bind_cols(ges_preds)
```

As a result of the above pipeline, you now have a tibble with predictions and the corresponding confidence interval.

```
ges_preds
```

```
# A tibble: 9 x 6
       fit se.fit  prob    LB    UB pinkie_curl
     <dbl>  <dbl> <dbl> <dbl> <dbl>       <int>
1    0.927  0.225 0.717 0.619 0.797           1
2    0.790  0.186 0.688 0.605 0.760           2
3    0.652  0.152 0.657 0.588 0.721           3
4    0.514  0.127 0.626 0.566 0.682           4
5    0.377  0.118 0.593 0.536 0.647           5
6    0.239  0.127 0.559 0.498 0.620           6
7    0.101  0.152 0.525 0.451 0.598           7
8   -0.0366 0.186 0.491 0.401 0.581           8
9   -0.174  0.225 0.457 0.351 0.566           9
```

This can be used as the basis for a plot of the predicted probabilities (Figure 12.7), which plots the 95% confidence intervals around each fitted value using `geom_errorbar()`. The following command also sets the *x*-axis ticks to the integer sequence 1 to 9 with `scale_x_continuous()`. In addition, `xlab()` and `ylab()` are used to tweak the axis labels.

```
ges_preds %>% ggplot(aes(x = pinkie_curl, y = prob)) +
  geom_point(size = 3) +
  geom_errorbar(aes(ymin = LB, ymax = UB), width = 0.5) +
  scale_x_continuous(breaks = 1:9) +
  xlab('Pinkie curl') +
  ylab('p(y = Shape)') +
  theme_minimal()
```

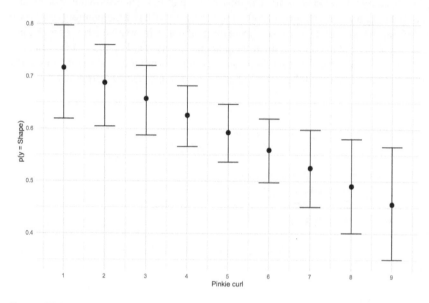

Figure 12.7. Predicted probability of observing a shape response as a function of a gesture's pinkie curl

12.7. Chapter Conclusions

With this chapter, the book turned away from ordinary linear models to generalized linear models, 'generalizing' the linear model framework to other data-generation processes. You learned about your first GLM: logistic regression.

Logistic regression assumes that the response has been generated by a process that is Bernoulli-distributed. The goal is to predict the parameter p, the probability of observing a particular event. The logistic function is used to make sure that regression doesn't predict impossible probabilities, crunching any number into the range [0,1] Then, you learned about the log odds, which is the internal 'metric' of logistic regression. The logistic function is used to transform log odds into probabilities.

The next chapter deals with another incredibly useful type of generalized linear model, Poisson regression. Mathematically, this model actually turns out to be easier than logistic regression! If you felt that this chapter was a little bit too much in terms of the math, please hang on and continue reading. The next chapter will also give a 'big picture' overview of the generalized linear model framework. This will help to clarify certain aspects of logistic regression.

12.8. Exercises

12.8.1. Exercise 1: Re-analysis of the Gesture Data with a Centered Predictor

For the gesture data, center the pinkie curl variable and rerun the analysis. How does the intercept change? What is its log odds value and what is the corresponding probability?

12.8.2. Exercise 2: Incorporate an Additional Predictor

For the gesture data, incorporate the additional predictor `index_curve`. This predictor quantifies the degree to which the index finger is curved. How does this variable affect the proportion of shape responses? Compare these results to contingency tables that compare the number of shape/height responses to the index curve variable.

12.8.3. Exercise 3: Incorporating a Nonlinear Effect

Chapter 8 introduced polynomial regression. It turns out that the effect of the pinkie curl variable in Hassemer and Winter (2016) shows some nonlinearities. After centering the pinkie curl variable, create a quadratic version of this variable and add it to the model. Is the quadratic effect significant? If so, what does the quadratic effect correspond to conceptually?

13 Generalized Linear Models 2

Poisson Regression

13.1. Motivating Poisson Regression

Poisson regression is another type of generalized linear model, and it's just as useful as logistic regression. However, interestingly, Poisson regression is surprisingly underutilized in the language sciences. This is perplexing because the Poisson distribution is the canonical distribution for count processes and, if linguists like to do one thing, it's counting! They like to count words, grammatical constructions, sociolinguistic variants, speech errors, discourse markers, and so on. In all of these cases, a Poisson model is the natural choice.

Since Poisson regression is not as common as logistic regression, let me tell you about a few cases where I have used this type of model. I hope that this will convince you of the utility of this type of generalized linear model. In Winter and Grawunder (2012), we used Poisson regression to model the frequency of fillers (such as *uh* and *oh*) and certain discourse markers as a function of politeness contexts. In Bentz and Winter (2013), we conducted a typological study for which a particular version of Poisson regression was used to model how many case markers a language has as a function of the proportion of second language learners. Here, Poisson regression allowed us to assess the impact of language contact on a language's morphological complexity. In Winter, Perlman, and Majid (2018), we used a version of Poisson regression to model the frequency of words as a function of their sensory modality, testing the idea that the English language is characterized by visual dominance. In fact, whenever word frequency is your response measure (and it often is in linguistics, especially in corpus linguistics), you should consider using a Poisson regression model.

Figure 13.1a shows an artificial dataset where speech error counts are related to blood alcohol concentration. Whereas speech errors were treated in a binary fashion in the last chapter (presence or absence of speech error), it is treated as a count variable here. The corresponding Poisson regression fit is shown as a thick line. This line can be interpreted to represent the mean rate of a speech error occurring. Notice how the speech error counts are more variable for high blood alcohol concentration than for low blood alcohol concentration, which is a form of heteroscedasticity (discussed below).

13.2. The Poisson Distribution

In the context of Poisson regression, the response y is assumed to be generated by a process that follows a Poisson distribution.

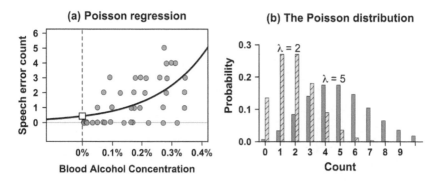

Figure 13.1. (a) Speech error count as a function of blood alcohol concentration with superimposed Poisson regression fit (bold curve); the white square represents the intercept; (b) two versions of the Poisson distribution with rates of 2 and 5

$$y \sim Poisson(\lambda) \tag{E13.1}$$

The Poisson distribution is shown in Figure 13.1b for two representative parameters, with the height of the bars indicating the probability of particular counts. The Poisson distribution only has one parameter, λ 'lambda', which specifies the rate of a count process. If lambda is high, then the rate of observing an event (such as speech errors, fillers, or grammatical markers) is high. Notice how, for the low rate $\lambda = 2$ (striped bars), the Poisson distribution indicates that the counts 1 and 2 are the most probable. Not observing an event (a count of 0) is slightly less probable, and so are counts much in excess of 2.

Importantly, the Poisson distribution is bounded by 0—counts cannot be negative. And the distribution is a discrete (categorical) distribution, in that only positive integers are possible, with no in-betweens.

Another peculiar property of the Poisson distribution is that the variance of the distribution is married to λ, in stark contrast to the normal distribution, where the standard deviation σ is an independent parameter that needs to be estimated. You can see this in Figure 13.1b by the fact that the distribution for $\lambda = 5$ has a higher spread than the distribution for $\lambda = 5$. For low rates, the variance is low because the distribution is bounded by 0, with no way of extending beyond that boundary. For high rates, the distribution can extend in both directions, towards lower and higher counts. You can think of this as 'heteroscedasticity' (unequal variance, see Chapters 4 and 6) as being built into this distribution.

Poisson regression models the parameter λ as a function of some predictors. The problem is that our familiar equation '$\beta_0 + \beta_1 * X$' can predict *any* value, but λ, being a rate parameter, can only be positive. Thus, you need a function that restricts the output of '$\beta_0 + \beta_1 * x_i'$' to positive values. The function that achieves this is the exponential function, as shown in Figure 13.2.

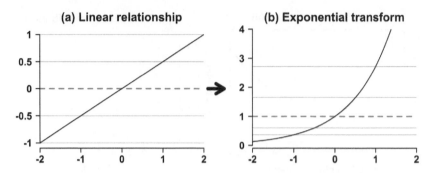

Figure 13.2. (a) A linear relationship between *y* and *x*; transforming the *y* s with the exponential function (b) restricts them to the positive range; the dashed line shows that, when 0 is exponentially transformed, it becomes 1

Thus, wrapping the exponential function around ' $\beta_0 + \beta_1 * x_i'$ ' will ensure that no negative values can be predicted.

$$\lambda_i = exp\left(\beta_0 + \beta_1 * x_i\right) \tag{E13.2}$$

Remember from the last chapter and Chapter 5 that the logarithm is the inverse of the exponential function. If we take the logarithm for both sides of E13.2, the equation becomes:

$$\log(\lambda_i) = \beta_0 + \beta_1 * x_i \tag{E13.3}$$

Thus, the familiar equation ' $\beta_0 + \beta_1 * x_i'$ ' will predict log-lambdas. Similar to logistic regression, this means that we will have to be careful in interpreting the output, for which this chapter will give a lot of guidance.

13.3. Analyzing Linguistic Diversity Using Poisson Regression

You are now going to analyze the data from Nettle's (1999) book on linguistic diversity, a dataset that was very briefly introduced in Chapters 1 and 2. To remind you of Nettle's hypothesis: countries with lower ecological risk (more fertile environments) are predicted to have higher linguistic diversity. Nettle (1999) analyzed the data by log-transforming language counts. This is sub-optimal (O'Hara & Kotze, 2010) because, amongst other things, count data will often exhibit heteroscedasticity even after log-transforming. The Poisson is the canonical distribution for counts.

Let's start by loading in the spreadsheet with `read_csv()`.

```
library(tidyverse)
library(broom)

nettle <- read_csv('nettle_1999_climate.csv')

nettle
```

```
# A tibble: 74 x 5
         Country Population   Area   MGS Langs
           <chr>      <dbl>  <dbl> <dbl> <int>
1        Algeria       4.41   6.38  6.60    18
2         Angola       4.01   6.10  6.22    42
3      Australia       4.24   6.89  6.00   234
4     Bangladesh       5.07   5.16  7.40    37
5          Benin       3.69   5.05  7.14    52
6        Bolivia       3.88   6.04  6.92    38
7       Botswana       3.13   5.76  4.60    27
8         Brazil       5.19   6.93  9.71   209
9   Burkina Faso       3.97   5.44  5.17    75
10           CAR       3.50   5.79  8.08    94
# ... with 64 more rows
```

The columns `Population` and `Area` contain the log_{10} population size and area of the country, respectively. For the time being, the relevant variables for our analysis will be `MGS` (mean growing season, a measure of ecological risk, the predictor) and `Langs` (the number of languages within a country). The `MGS` predictor indicates how many months per year one can grow crops in a country.

As always, it makes sense to first get a feel for this dataset, for example, by checking the range of the `MGS` variable:

```
range(nettle$MGS)
```

```
[1] 0 12
```

This shows that there are some countries in which one cannot grow crops at all (MGS = 0 months), as well as others where one can grow crops throughout the entire year (MGS = 12 months). The following code displays those countries that have a mean growing season of either 0 or 12.

```
filter(nettle, MGS == 0 | MGS == 12)
```

```
# A tibble: 6 x 5
  Country    Population   Area   MGS Langs
    <chr>         <dbl>  <dbl> <dbl> <int>
1 Guyana         2.90    5.33  12.     14
2 Oman           3.19    5.33   0.      8
```

```
3 Solomon Islands        3.52  4.46  12.   66
4 Suriname               2.63  5.21  12.   17
5 Vanuatu                2.21  4.09  12.  111
6 Yemen                  4.09  5.72   0.    6
```

```
# same as:

filter(nettle, MGS %in% range(MGS))
```

The countries Guyana, Solomon Islands, Suriname, and Vanuatu have a mean growing season of 12, indicating minimal ecological risk (fertile environments that facilitate local subsistence farming). In contrast, Oman and Yemen have a mean growing season of 0, indicating maximal ecological risk (arid environments that make people form widespread trade networks). Notice, furthermore, how Oman and Yemen have fewer distinct languages (Langs) than the other countries, which is our first indication that Nettle's hypothesis might be right.

To model linguistic diversity as a function of ecological risk with Poisson regression, glm() is the relevant function. As was the case in the last chapter, the argument family is used to specify the type of GLM. This time around, specify family to be 'poisson'.

```
MGS_mdl <- glm(Langs ~ MGS, data = nettle,
               family = 'poisson')

tidy(MGS_mdl)
```

```
          term   estimate    std.error  statistic     p.value
1  (Intercept) 3.4162953  0.039223267   87.09869  0.000000e+00
2          MGS 0.1411044  0.004526387   31.17375  2.417883e-213
```

The coefficients of a Poisson model are represented as logarithms. Thus, exponentiation will be needed to report the predicted mean rate. Let's perform some example calculations. First, extract the coefficients.

```
mycoefs <- tidy(MGS_mdl)$estimate

# Extract intercept and slope:

intercept <- mycoefs[1]

slope <- mycoefs[2]

# Check:

intercept
```

```
[1] 3.416295
```

```
slope
```

```
[1] 0.1411044
```

Let's see what the model predicts for the full range of MGS values from 0 to 12 months.

```
intercept + 0:12 * slope
```

```
[1] 3.416295 3.557400 3.698504 3.839609 3.980713 4.121818
[7] 4.262922 4.404026 4.545131 4.686235 4.827340 4.968444
[13] 5.109549
```

So, for a mean growing season of 0 months, the model predicts a log language rate of 3.41. For a 1-month growing season, the model predicts a log rate of 3.56, and so on. Exponentiating these fitted values yields the estimated lambdas.

```
exp(intercept + 0:12 * slope)
```

```
 [1]  30.45637  35.07188  40.38685  46.50727  53.55521
 [6]  61.67123  71.01719  81.77948  94.17275 108.44415
[11] 124.87831 143.80298 165.59559
```

These can meaningfully be interpreted as the mean rate of languages. That is, you expect a country with 0 months MGS to have about 30 languages. On the other hand, a country with 12 months MGS is predicted to have about 166 languages.

For plotting purposes, let's create a more fine-grained sequence of predictions from 0 to 12 in a step-size of 0.01.

```
myMGS <- seq(0, 12, 0.01)
```

Let's generate predictions with `predict()`, for which we need the MGS predictor to be in a tibble.

```
# Tibble to generate predictions for:

newdata <- tibble(MGS = myMGS)

newdata
```

```
# A tibble: 1,201 x 1
    MGS
  <dbl>
1  0
2  0.01
3  0.02
4  0.03
```

```
 5   0.04
 6   0.05
 7   0.06
 8   0.07
 9   0.08
10   0.09
# ... with 1,191 more rows
```

The log predictions can be computed as follows:

```
MGS_preds <- predict(MGS_mdl, newdata)

head(MGS_preds)

       1        2        3        4        5        6
3.416295 3.417706 3.419117 3.420528 3.421939 3.423350
```

Exponentiating this yields the estimates of λ, the rate of language occurrence as a function of growing season.

```
MGS_preds <- exp(MGS_preds)

head(MGS_preds)

       1        2        3        4        5        6
30.45637 30.49938 30.54245 30.58557 30.62876 30.67201
```

Alternatively, you can use the `type = 'response'` to make the `predict()` function compute the estimated lambdas directly.

```
MGS_preds <- predict(MGS_mdl, newdata, type = 'response')

head(MGS_preds)

       1        2        3        4        5        6
30.45637 30.49938 30.54245 30.58557 30.62876 30.67201
```

Next, put everything into a tibble for `ggplot2`.

```
mydf <- tibble(MGS = myMGS, Rate = MGS_preds)

# A tibble: 1,201 x 2

    MGS  Rate
  <dbl> <dbl>
1 0      30.5
2 0.01   30.5
3 0.02   30.5
4 0.03   30.6
5 0.04   30.6
6 0.05   30.7
```

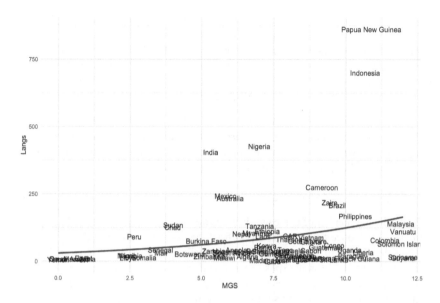

Figure 13.3. Linguistic diversity as a function of mean growing season (Nettle, 1999) with Poisson regression fit; the line represents the predicted language rate

```
 7   0.06   30.7
 8   0.07   30.8
 9   0.08   30.8
10   0.09   30.8
# ... with 1,191 more rows
```

You can then plot the data with a superimposed Poisson regression fit, as shown in Figure 13.3. In this code chunk, `geom_line()` draws from the tibble with predictions, and it gets a new set of mappings.

```
nettle %>% ggplot(aes(x = MGS, y = Langs)) +
  geom_text(aes(label = Country)) +
  geom_line(data = mydf, mapping = aes(x = MGS, y = Rate),
            col = 'blue', size = 1) +
  theme_minimal()
```

13.4. Adding Exposure Variables

Nettle (1999) observed that, in order for the regression of languages on mean growing season to be meaningful, one has to control for a country's size. Obviously, larger countries tend to have more different languages, such as India, which is a very big country that also has a lot of linguistic diversity. In this case, a country's area determines what in Poisson regression is called 'exposure': more area means more opportunities for observing high counts. You can adjust a rate by an exposure variable, which is area in this case, but it could be time in other applications. For example, if you were to conduct an experiment where you are counting speech errors in trials with varying durations, there are naturally going to be higher counts for longer trials.

For exposure variables, the rate lambda is split into two components, the mean number of events μ, per unit of exposure τ 'tau'. For example, μ could be the average number of languages per country, and τ could be the size of the country. Alternatively, μ could be the average number of speech errors, and τ could be the duration of a trial. The rate then is $\lambda = \frac{\mu}{\tau}$, such as 'number of languages per square mile' or 'number of speech errors per second'. I will spare you the derivation of the following formula, but in the presence of an exposure variable, the equation for the predicted log number of languages is the following.[1]

$$\log(\mu) = \beta_0 + \beta_1 \times MGS + \log(\tau) \tag{E13.4}$$

Thus, the log number of languages is a function of $'\beta_0 + \beta_1 \times MGS'$ and the exposure term $\log(\tau)$. Notice that there's no β in front the exposure variable—this term has no coefficient, which also means that nothing is actually estimated for this term. In this case, adding a country's area as an exposure term is effectively saying that the average number of languages occurring in a country is directly proportional to the size of a country.

In R, all of this is easy. You simply need to wrap `offset()` around the exposure variable of interest—in this case, `Area`.

```
MGS_mdl_exposure <- glm(Langs ~ MGS + offset(Area),
                        data = nettle, family = 'poisson')

tidy(MGS_mdl_exposure)
         term    estimate   std.error  statistic  p.value
1 (Intercept) -2.8230092 0.040738134  -69.29648         0
2         MGS  0.2092749 0.004719774   44.34003         0
```

Notice how, compared to the model without the exposure variable, the slope of MGS variable has increased by about 50%. Thus, after controlling for a country's size, the relationship between ecological risk and linguistic diversity is estimated to be even stronger.

---

1 If you're mathematically inclined, read this footnote. First, the logarithm of a quotient such as $\frac{\mu}{\tau}$

can be expressed as a subtraction of two logarithms, so:

$$\log(\lambda) = \log\left(\frac{\mu}{\tau}\right) = \log(\mu) - \log(\tau)$$

Combining this with the predictive equation yields:

$$\log(\mu) - \log(\tau) = \beta_0 + \beta_1 \times MGS$$

Next, move the exposure term $\log(\tau)$ over to the right-hand side of the equation, which yields E13.4. McElreath (2016: 312–313) has a particularly clear discussion of this.

To give another example of the usefulness of exposure variables, consider the above-mentioned study on politeness markers (Winter & Grawunder, 2012). As mentioned before, this study investigated the rate with which Korean speakers used fillers such as *uh* and *oh* as a function of politeness contexts. Crucially, our task was open-ended, which meant that, for each trial, participants could talk as much as they wanted. This resulted in some fairly long utterances, as well as some very short ones. Naturally, longer utterances contain more fillers, which we dealt with by adding the exposure variable 'utterance length'.

13.5. Negative Binomial Regression for Overdispersed Count Data

As discussed above, the variance of the Poisson distribution scales with the mean: the higher the mean rate, the more variable the counts. However, it is possible that in an actual dataset, the variance is larger than is theoretically expected for a given lambda. If this happens, you are dealing with what's called 'overdispersion' or 'excess variance'.[2]

You can compensate for overdispersion by using a variant of Poisson regression that is called 'negative binomial regression'. Essentially, this is a generalization of Poisson regression where the variance is 'freed' from the mean. In other words, the constraint that the mean is equal to the variance is relaxed for negative binomial regression. Other than that, everything else that you've learned about Poisson regression stays the same.

Let us refit the above model (with exposure variables), this time using negative binomial regression rather than Poisson regression. For this, the `glm.nb()` function from the `MASS` package can be used (Venables & Ripley, 2002).

```
library(MASS)

# Fit negative binomial regression:

MGS_mdl_nb <- glm.nb(Langs ~ MGS + offset(Area),
                     data = nettle)

tidy(MGS_mdl_nb)
```

	term	estimate	std.error	statistic	p.value
1	(Intercept)	-3.0527417	0.26388398	-11.568500	5.951432e-31
2	MGS	0.2296025	0.03418441	6.716585	1.860333e-11

First, notice that the standard error for the MGS slope has increased by quite a bit compared to the corresponding Poisson model. Negative binomial models are generally the more conservative option if there actually is overdispersion present in the data. Let's have a look at the `summary()` output of the model.

---

2 The term 'underdispersion' describes cases where there is less variance than theoretically expected under the Poisson distribution. I have never encountered underdispersion in a linguistic dataset up to this point.

```
summary(MGS_mdl_nb)
```

```
Call:
glm.nb(formula = Langs ~ MGS + offset(Area), data = net-
tle, init.theta = 1.243938533,
    link = log)


Deviance Residuals:
    Min       1Q    Median       3Q       Max
-2.3904   -0.9479   -0.4620    0.2822    2.5034


Coefficients:
             Estimate Std. Error z value Pr(>|z|)
(Intercept) -3.05274    0.26388 -11.568  < 2e-16 ***
MGS          0.22960    0.03418   6.717 1.86e-11 ***
---
Signif. codes:  0 '***' 0.001 '**' 0.01 '*' 0.05 '.' 0.1 ' ' 1


(Dispersion parameter for Negative Binomial(1.2439) family
taken to be 1)


    Null deviance: 120.21  on 73   degrees of freedom
Residual deviance:  82.25  on 72   degrees of freedom
AIC: 771.88


Number of Fisher Scoring iterations: 1
              Theta:  1.244
          Std. Err.:  0.190


2 x log-likelihood: -765.87
```

You can test whether there is a 'significant' degree of overdispersion via overdispersion tests, one implementation of which is the odTest() function from the pscl package (Jackman, 2015). This function performs a likelihood ratio test (see Chapter 15), comparing the likelihood of the negative binomial model against the likelihood of the corresponding Poisson model.

```
library(pscl)

# Perform overdispersion test:

odTest(MGS_mdl_nb)
```

```
Likelihood ratio test of H0: Poisson, as restricted NB
model:
n.b., the distribution of the test-statistic under H0 is
non-standard
e.g., see help(odTest) for details/references
```

```
Critical value of test statistic at the alpha= 0.05 level:
2.7055
Chi-Square Test Statistic = 5533.0321 p-value = < 2.2e-16
```

In this case, the difference in likelihood between the two models is significant (p-value = < 2.2e-16), indicating that you should use a negative binomial model rather than simple Poisson regression. If you wanted to report the results of this overdispersion test, I would write something like this: 'A likelihood ratio test of a negative binomial model against a Poisson model revealed a significant difference $(\chi^2(1) = 5533.03, p < 0.0001)$.' The '1' inside the brackets represents the degrees of freedom, which, as was the case with the model comparisons performed in Chapter 11, indicates the difference in the number of estimated parameters for the two models. The number is 1 because negative binomial regression estimates one additional parameter (the dispersion parameter). The test statistic for this is χ^2 'chi squared'—this is a new test statistic, just like t and F discussed in previous chapters. I am not going to go into the details of the chi-square distribution here.

13.6. Overview and Summary of the Generalized Linear Model Framework

Figure 13.4 summarizes the aspects of the generalized linear model framework that you have learned about so far.

There's one new bit that you haven't been exposed to so far, which is the function $I()$ wrapped around the predictor for linear regression. This is what is called the 'identity function'. The term 'identity function' simply is math-speak for a function that preserves identity. Or, in other words, this function does nothing: $I(x) = x$. The identity function in this case can be paraphrased as follows: 'Take the predictor *as is*, with no transformations.' The reason for adding it to Figure 13.4 is that it shows the parallelism between the different types of models. It allows you to see that linear regression is actually a specific case of the generalized linear model, namely, a GLM where the output of the predictive equation *isn't* transformed. Thus, the GLM framework subsumes linear regression.

Figure 13.4 furthermore highlights that each GLM has three components. First, a distribution for the data-generating process. Second, a predictive equation, what is called the 'linear predictor' in the GLM framework, and this is what you are used to thinking of as your model's equation. The third component is the 'link function', which *links* the linear predictor to the parameter of interest. This function ensures that the linear predictor predicts sensible values for each parameter, that is, values between 0 and 1 for p, and positive values for λ. Perhaps somewhat confusingly, the link functions are named after their inverses: logistic regression uses the logit or 'log odd' link, Poisson regression uses the

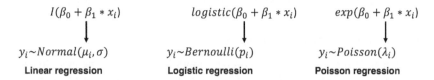

Figure 13.4. Summary of the three types of generalized linear models covered so far

log link function. As a result of the link function, logistic regression returns log odd predictions, and Poisson regression returns log predictions. For logistic regression, you need the logistic function to transform the log odd predictions into probabilities. For Poisson regression, you need the exponential function to transform the log predictions into rates.

13.7. Chapter Conclusions

In this chapter, you have extended your knowledge of generalized linear models. Specifically, you have learned how to model count data with Poisson regression, and its extension, negative binomial regression. The coefficients of a Poisson model are shown as log coefficients, which means that, after calculating the log predictions, you need to use exponentiation to interpret your model in terms of average rates. To control for differential exposure (such as countries of varying sizes, time intervals of varying durations, etc.), exposure variables can be added. Negative binomial regression was used to account for overdispersion. The set of tools learned in this chapter allow modeling a wide array of data structures. Whenever you see count data, Poisson regression and its sister, negative binomial regression, should be your go-to tools.

Finally, this chapter concluded with an overview of the generalized linear model framework. In particular, it was highlighted that each GLM has three components: a distribution for the data-generating process, a linear predictor, and a link function.

13.8. Exercises

13.8.1. Exercise 1: Getting a Feel for Poisson Data

The `rpois()` function can be used to generate random data that is Poisson-distributed. Notice how all the random number generation functions in R start with the letter 'r', which stands for 'random'. You already know `rnorm()` and `runif()`.

You supply two things to the `rpois()` function: lambda, and how many numbers you want to generate.

```
rpois(50, lambda = 2)   # output not shown
```

You can tabulate the counts and plot them as follows:

```
plot(table(rpois(50, lambda = 2)))
```

Play around with different lambdas to get a feel for the Poisson distribution.

13.8.2. Exercise 2: Visual Dominance

Winter et al. (2018) showed that, on average, English visual words are more frequent than words for the other modalities. This exercise asks you to retrace this analysis focusing on the subset of adjectives (the paper also included verbs and nouns). First, load in the Lynott and Connell (2009) sensory modality ratings, as well as the English Lexicon Project data file which contains SUBTLEX word frequencies.

```
lyn <- read_csv('lynott_connell_2009_modality.csv')
ELP <- read_csv('ELP_full_length_frequency.csv')
```

Next, merge the information from the ELP tibble into the lyn tibble:

```
both <- left_join(lyn, ELP)
```

Select only the relevant columns. If you have the MASS package loaded from the previous exercises, there will be a naming conflict, as this package also contains a function called select(), just like dplyr. The following command tells R that you mean the dplyr function, not the MASS function:

```
both <- dplyr::select(both,
    Word, DominantModality:Smell, Log10Freq)
```

Finally, to apply Poisson regression, you need the frequency variable as positive integers.

```
both <- mutate(both, Freq = 10 ^ Log10Freq)
```

Next, fit a model Poisson regression model with Taste, Smell, Touch, Sight, and Sound as predictors (all of these are continuous rating scales). After this, fit a negative binomial regression model using the glm.nb() function from the MASS package. Check whether there is significant overdispersion with the odTest() function from the pscl package. Interpret the Poisson and negative binomial regression outputs. Do English speakers use visual adjectives more frequently? What about smell adjectives in comparison? How do the results of the Poisson and negative binomial regression compare to each other?

As an additional exercise, you may want to assess collinearity between the different predictors (see Chapter 6) with the vif() function from the car package.

14 Mixed Models 1
Conceptual Introduction

14.1. Introduction

As you progressed through this book, the range of datasets that you could model has continuously increased. This chapter will continue this trend, introducing you to mixed effects models, which are also known as multilevel models.[1] These models are very common in the language sciences and related disciplines. Many datasets require some form of mixed model. Luckily, everything that you have learned up to this point is still relevant. As mixed models are an extension of regression, a lot will look familiar to you. This chapter covers the conceptual side of mixed models. The next chapter covers implementation.

14.2. The Independence Assumption

You have been exposed to what are commonly called the 'normality assumption' and the 'constant variance assumption' in various chapters (Chapter 4; see also Chapters 6 and 12). The importance of these two assumptions is *far* outweighed by the 'independence assumption'. As you will see below, if the independence assumption is violated, the results of statistical tests cannot be trusted.

What is independence? Rolling a die repeatedly is a nice example of a truly independent process. Granted that you shake the die thoroughly before rolling it, the outcome of each roll is independent from another one. A dependence, then, is any form of connection between data points. For example, two data points could be connected by virtue of coming from the same participant. In that case, these data points are not independent anymore. Most of the time, multiple data points from the same participant are more similar to each other than data points from different participants. You can think of this as a statement about the residuals: if participant A performs overall differently from participant B, then all of participant A's residuals will act as a group, and so will all of participant B's residuals.

Violations of the independence assumption have *massive* effects on the Type I error (false positive) rates of a study. In Winter (2011), I performed a simple simulation to demonstrate the detrimental effects of violating the independence assumption. I used speech production research as an example, where it is common practice to include

---

1 You may also hear the term 'hierarchical linear model', which is often used to refer to mixed effects models with a nested hierarchical structure (e.g., "pupil within classroom within school").

exact repetitions of linguistic items; that is, the same word or sentence is uttered multiple times by the same speaker (for a critical discussion of this practice, see Winter, 2015). These repetitions introduce dependencies into one's datasets, as each repetition is always from the same participant, as well as from the same item. Adding lots of repetitions to your experimental design and not telling your model about this amounts to artificially inflating the sample size. My simulations showed that if these dependency structures are not accounted for in one's models, Type I error rates quickly climb way past the commonly accepted 0.05 threshold.

Similar detrimental effects of violating the independence assumption have been discussed extensively in other fields under the banner of 'pseudoreplication' and the 'pooling fallacy' (Hurlbert, 1984; Machlis, Dodd, & Fentress, 1985; Kroodsma, 1989, 1990; Lombardi & Hurlbert, 1996; Milinski, 1997; García-Berthou & Hurlbert, 1999; Freeberg & Lucas, 2009; Lazic, 2010).

All sorts of linguistic datasets include non-independent cases. For example, nearly every psycholinguistic, phonetic, or sociolinguistic study uses a 'repeated measures design' where multiple data points are collected from the same participant. Dependencies between data points are also abundant in typological studies. For example, due to them all coming from the same language family, one cannot treat German, English, Farsi, and Hindi as independent data points in a statistical analysis (Jaeger, Graff, Croft, & Pontillo, 2011; Roberts & Winters, 2013). Finally, non-independences are also abundant in corpus linguistics, where there are often multiple data points from the same text, or the same author, or the same newspaper or publishing house, etc.[2]

14.3. Dealing with Non-independence via Experimental Design and Averaging

How can one deal with violations of the independence assumption? Whether or not the independence assumption has been violated is something that has to do with the design of a study, as well as with how a study is analyzed. One can deal with the independence assumption by designing studies that minimize dependence between data points. For example, in some circumstances it may be possible to perform single-trial between-participant experiments where each participant only contributes one data point. In fact, you have seen data coming from such experiments throughout this book. For example, the similarity-is-proximity study by Winter and Matlock (2013) discussed in Chapter 8 was of this sort: each participant was exposed to a condition where they either read the similar or the dissimilar text. No participant in this study was exposed to both conditions. Another example of a single-trial between-participants experiment was the gesture perception study by Hassemer and Winter (2016) discussed in Chapter 12. Here, each participant saw only one of the hand shapes from the 3D hand shape continuum. In all of these cases, it was possible to fit simple linear models or generalized linear models without violating the independence assumption

---

2 In corpus linguistics, the independence assumption is routinely ignored by researchers. For example, chi-square tests (see Appendix A) are performed on tables with multiple data points from the same text or author. Gries (2015) says that mixed models are the 'most under-used statistical method in corpus linguistics'.

because the experimental design ensured that there weren't multiple data points from the same participant.

Another way of dealing with non-independences is via aggregation. If you had multiple data points from the same participant, why not average everything so that each participant only contributes one data point? This is a possible way of dealing with non-independent cases, but it's not the optimal way, because whenever you compute averages you lose information (see Chapter 4). In particular, the *variation* across the non-independent cases is not retained in the final analysis. If your statistical model only 'sees' the averaged data points, it is going to underestimate the amount of variation present in the data. This also means that the model has less information available for making adequate inferences.

In fact, you've performed several analyses on averaged values throughout this book. In studies that use 'norms' (psycholinguistic ratings), it is routinely the case that each word is associated with a value that comes from averaging over the ratings of multiple participants. Such averaging characterized the Warriner et al. (2013) emotional valence data (discussed in Chapters 3 and 7) and the iconicity rating data (discussed in Chapters 6 and 8). In these analyses, each word only contributed one data point, thus assuring independence across words. However, by-participant variability in the ratings is not retained in the final analysis if you are working with by-word averages.

14.4. Mixed Models: Varying Intercepts and Varying Slopes

So, it's a good idea to avoid averaging whenever this is an option. This is where mixed models come into play. These models allow incorporating non-independent clusters of data into one's analysis. In other words: you can tell your mixed model about the dependency structures within a dataset so that it makes appropriate estimates and draws appropriate inferences.

The primary workhorse for dealing with clusters of non-independent data points are what many researchers call 'random effects'; specifically, 'random intercepts' and 'random slopes'. Here, I will adopt the terminology of 'random effects', but I will use the terms 'varying intercepts' and 'varying slopes' instead of 'random intercepts' and 'random slopes'. These concepts will be explained via a discussion of Figure 14.1.

Figure 14.1 depicts the relationship between response durations and trial order in a psycholinguistic experiment. The trial order variable on the *x*-axis indicates progression through the experiment (to the left: beginning of the experiment; to the right: end of the experiment). The data is shown separately for three participants, Yasmeen, Logan, and Dan. When participants progress through an experiment, they tend to either speed up or slow down. In this case, Yasmeen and Logan speed up (negative slopes: shorter durations for later trials), and Dan slows down (positive slope: longer durations for later trials). In addition, participants differ in whether they are overall faster (particularly Yasmeen) or slower (particularly Dan).

In Figure 14.1, the population-level estimate is always shown as a solid black line. This is what the mixed model predicts as the average effect across all participants. The dashed lines indicate the participant-specific estimates. The top row indicates the fit of a mixed model that allows participants to have varying intercepts, but not varying slopes. As a result, this model allows the lines to differ in their intercepts (they are shifted upwards or downwards) but not their slopes (the dashed lines are forced to be

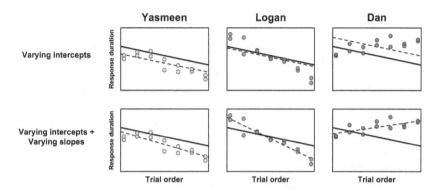

Figure 14.1. Response durations as a function of trial order ('experiment time') for three different participants; Yasmeen and Logan speed up throughout the experiment, Dan slows down; the bold line shows the population-level estimates (across participants); the dashed lines represent the participant-specific 'random effect' estimates; top row: a model with varying intercepts but not varying slopes tends to mischaracterize the participant-specific response patterns; bottom row: a model with varying intercepts and varying slopes characterizes the participant-specific trends more accurately

parallel to the population line). Because the model hasn't been instructed to estimate varying slopes, it is quite off for some participants—in this case, particularly for Dan. The fact that Dan slows down cannot be captured in a model where all slopes are restricted to be parallel.

The bottom row indicates the fit of a mixed model that allows participants to have varying intercepts, as well as varying slopes. This time around, the dashed lines are much closer to the data of each participant. That is, the model adequately captures the fact that some participants differ in their trial order effect.

The following equations show a simplified representation of mixed models. The predictor 'trial' in these equations represents how far along a participant is in the experiment (the x-axis in Figure 14.1).

Varying intercept model ('random intercept model'):

$$y = \beta_{0j} + \beta_1 * trial + \varepsilon \tag{E14.1}$$

Varying intercept, varying slope model ('random slope model'):

$$y = \beta_{0j} + \beta_{1j} * trial + \varepsilon \tag{E14.2}$$

In this equation, y is conditioned on the trial variable, with an intercept (β_0) and a slope (β_1) for the trial effect. This is something that you are all too familiar with by now. Notice one extra bit, however: in equation E14.1, the intercept β_0 bears the subindex j. This subindex stands for 'participant' in this case. $j=1$ is one specific participant

(Yasmeen), $j=2$ is another participant (Logan), and so on. The fact that the intercept bears this subindex means that different participants have different intercepts. In other words, there are different β_0s for different js.

In the second model (E14.2), *both* the intercept and the slope bear the subindex, which means that both coefficients differ between participants. Specifically, each participant now also gets their own specific slope estimate. For example, for $j=1$ (Yasmeen), the intercept β_0 is 800ms and the slope β_1 is –50. For $j=2$ (Logan), the intercept β_0 is 1000ms and the slope β_1 is –90.

When people use the term 'varying slopes' or 'random slopes' model, this generally subsumes 'varying intercepts/random intercepts'. That is, it is assumed that, when you let slopes vary, you also let intercepts vary.

Random effects are commonly contrasted with 'fixed effects'. There's nothing special about fixed effects; these are just the predictors that you have dealt with all throughout this book. In fact, throughout all chapters, you have been fitting fixed-effects-only models. The only thing that changes when turning to mixed models is that you can allow the relationship between y and a fixed effects predictor x to vary by individual (varying slopes). Mixed models get their name from the fact that they mix 'fixed effects' and 'random effects'.

What are some examples of common random effects in linguistics? In many areas of linguistics, particularly psycholinguistics and phonetics, 'participant' and 'item' are common random effects (see Baayen, Davidson, & Bates, 2008). An 'item' could be anything from a visual stimulus in a picture-naming study to a sentence in a study on sentence comprehension. The key thing here is that if there are multiple data points for the same item, this introduces a dependency into your dataset that needs to be modeled. In typology, common random effects include 'language family' and 'language contact area' (Jaeger et al., 2011; Bentz & Winter, 2013; Roberts & Winters, 2013). Unfortunately, mixed models aren't generally used in corpus linguistics, even though they are certainly necessary here as well, such as when there are multiple data points from the same author, text, newspaper, or publishing outlet. Each of these grouping factors could be a viable random effect in a corpus linguistic study.

Fixed effects are assumed to be constant across experiments. In this sense, they are repeatable. You could, for example, repeat a study on gender differences by collecting data with new female and male participants. While the individual participants vary (and their individual differences are a source of 'random' influence on the data), the gender effect can be tested again and again with new samples. The same way, you could repeat a study on sensory modality differences (e.g., taste versus smell) by selecting new words that differ along this dimension. The effects of 'gender' and 'modality' thus qualify as fixed effects because they are assumed to have a predictable, non-idiosyncratic influence on the response that could be tested with new samples of speakers or words.

Notice one more detail. While fixed effects can be continuous (Chapters 4–6) or categorical (Chapter 7), random effects are necessarily categorical. Why? Think about it this way: the whole point of fitting a mixed model is to account for dependent clusters of data points that somehow group together. The concept of a 'group' is a categorical one. In the above example, the data points from Yasmeen stick together as a group, and they are different from the data points of both Logan and Dan. The levels 'Yasmeen', 'Logan', and 'Dan' are part of a categorical factor 'participant' in this case. To see why random effects have to be categorical, it may also help to think from a sampling perspective: sampling from a population, such as the population of all speakers, involves sampling discrete units of that population.

14.5. More on Varying Intercepts and Varying Slopes

Let's delve more deeply into the topic of varying intercepts and varying slopes. We'll stick with the example of trial order, as discussed in the context of Figure 14.1. To make this example look more realistic, let's increase the sample size. I simulated data from 40 different participants, each of which provided responses for 12 trials, which yields a total of 480 different data points.

Figure 14.2a shows the fit of a simple linear regression model to this data, with no random effects. This model has only one error term, which is represented by the histogram of the residuals at the bottom of Figure 14.2a. The problem with this model is that it doesn't 'know' that many of these residuals are connected. In ignoring the fact that the data comes from only 40 participants, the model treats each data point as independent.

Figure 14.2b shows a model with random effects, specifically varying intercepts. The lines for the individual participants are shifted upwards and downwards, but they are all parallel to the average estimate (thick black line). Each participant is allowed to have their own intercept. You can think of this as assigning each participant a deviation score which describes how much that person's intercept deviates from the population intercept. A positive deviation score for a particular individual shifts the entire line for that individual upwards. A negative deviation score shifts it downwards. The deviation scores are shown in the histogram at the bottom. Notice the 'N' written at the bottom of this histogram, which indicates that this is a histogram of only 40 deviation scores, in contrast to the overall residuals, which are as many as there are data points in this dataset ($N = 480$) .

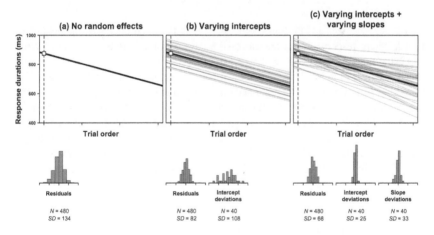

Figure 14.2. The relationship between trial order and response durations for 480 data points from 40 participants, modeled with (a) no random effects; (b) random effects: varying intercepts; and (c) random effects: varying intercepts plus varying slopes; gray lines correspond to individual participants

The problem with the model shown in Figure 14.2b is that it is inappropriate to assume that all participant slopes are the same, which amounts to saying that there are no individual differences for the trial order effect. Freeing the slopes from the constraint to be parallel to the population line yields an additional 40 deviation scores (Figure 14.2c). This time, the deviation scores describe deviations from the average slope. Say the average trial effect across all participants was −10 and one particular participant had an even steeper slope of −20. This participant would have a deviation score of −10, representing how much this specific participant's slope has to be adjusted downwards from the overall slope. Conceptually, at least, you can think of this as fitting individual regression lines for each participant.

There are a few more things to discuss in relation to Figure 14.2. First, it has to be mentioned that mixed models do not actually estimate one parameter per participant in this dataset. Instead, mixed models estimate the *variation around the specified random effect*. So, if you allow intercepts to vary by participants in your model, then this only adds one term to your model. The parameter estimated by this term is a standard deviation that represents the by-participant variation around the overall intercept. If you then additionally allow slopes to vary by participants, this adds another term. The number of histograms below each sub-plot shows you how many variance terms are being estimated for the models shown in Figure 14.2. Adding more participants would not add more histograms.

14.6. Interpreting Random Effects and Random Effect Correlations

The most widely used R package for mixed models in the language sciences is lme4 (Bates, Maechler, Bolker, & Walker, 2015). Have a look at the following lme4 model formula, in which RT is modeled as a function of the fixed effect trial.

```
RT ~ trial + (1 + trial|participant)
```

The 'RT ~ trial' bit looks familiar. This is the fixed effects component of this model, where response times are modeled as a function of the trial predictor. What's new is the random effect '(1 + trial|participant)'. In lme4 syntax, random effects are always in brackets. The vertical bar '|' bar inside the brackets can be paraphrased in plain English as 'conditioned on' or 'relative with respect to'. So, the expression '(1|participant)' would instruct lme4 to estimate by-participant varying intercepts because '1' acts as a placeholder for the intercept (see Chapter 4.8). The expression '(1 + trial|participant)' instructs the model to estimate by-participant varying intercepts, as well as by-participant varying slopes. Fitting a linear mixed effects model with this specification, you might get an output that looks like this (abbreviated):

```
Random effects:
 Groups      Name        Variance Std.Dev. Corr
 participant (Intercept) 1276.8   35.73
             trial         10.5    3.24    -0.38
 Residual                3678.0   60.65
Number of obs: 6000, groups:  participant, 60
```

```
Fixed effects:
            Estimate  Std. Error  t value
(Intercept)  998.6023     4.8753   204.83
trial         -9.9191     0.4192   -23.66
```

The top bit shows the estimates for the random effects component of the model; the bottom bit shows the estimates for the fixed effects. The bottom table is what you are used to seeing as the coefficients table in your linear model output. Let's focus on this first. In this case, response durations become shorter by about 10ms for each additional trial. In other words: across the board, participants appear to speed up throughout the experiment. Perhaps this is because they are learning to adapt to the task. Let's also observe the fact that the intercept in the fixed effects component is about 1000ms, which is the predicted response duration for the zeroth trial.

Let's now focus on the random effects. Each standard deviation represented in the output corresponds to one random effects parameter that is estimated by the model. First, there is a standard deviation for by-participant varying intercepts. This standard deviation is about 36ms, which describes the by-participant variation around the average intercept (1000ms). You can apply the 68%–95% rule from Chapter 3 to gauge your intuition about these numbers. Given this rule, you expect 68% of the participant intercepts to lie within the interval 964ms and 1036ms. The 95% interval spans from 928ms and 1072ms (the intercept plus and minus two standard deviations). The same calculation can be done to gauge your intuition about the varying slopes. The standard deviation of the by-participant varying slopes is about 3. Since the slope is about –10, you expect about 68% of all participants to have slopes between –13 and –7. Conversely, you expect 95% of the slopes to lie between –16 and –4 (the slope plus and minus two standard deviations).

Finally, the output additionally lists a correlation for the random effects, specifically a varying intercept/varying slope correlation term. This is actually an additional parameter that is estimated. The fact that the estimated correlation between varying intercepts and varying slopes is –0.38 indicates that higher intercepts had lower trial slopes. This could be because those who start out slow at the beginning of the experiment (high intercepts) have more opportunity to speed up (steeper slopes).

Intercept/slope correlations are not to be ignored. They contain useful information about your data and can be theoretically interesting in many cases. Imagine you performed an experiment where you measured participants' accuracy as a function of an interference task (a distractor task that interferes with cognitive functioning). If the task is too easy, the interference manipulation may have little effect. It could happen that only the participants who found the task difficult to begin with show a strong interference effect. Figure 14.3a shows a hypothetical dataset that exemplifies this situation—a psycholinguist may describe this as a 'ceiling effect', because participants perform 'at ceiling', leaving little room for the condition effect to shine.

In such a situation, it could be the case that low accuracy intercepts are associated with more extreme interference effects. Taking the intercepts from Figure 14.3a (the values in the 'control' condition) and plotting them against the slopes yields

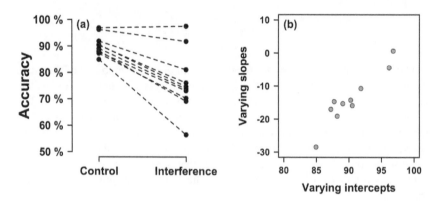

Figure 14.3. (a) Accuracy as a function of interference; each line represents a particular participant; higher-performing participants exhibit interference effects that are less steep; (b) plotting the varying intercepts against the varying slopes shows a correlation; each data point represents the estimates from one participant

Figure 14.3b, which shows a correlation of these two random effects. Participants with higher intercepts (better performers) had interference effects closer to 0.

14.7. Specifying Mixed Effects Models: `lme4` syntax

Now that you know about varying intercept/varying slope correlations, I can give you a more complete overview of the `lme4` notation for the different kinds of random effects structures that are commonly used in the language sciences, as shown in Table 14.1. You as the user of mixed models decide whether you want to fit a random intercept model (first row) or a random slope model (second row). In addition, you have control about whether you want to estimate the intercept/slope correlation term (second row) or not (third row).

As was mentioned before, what's to the left of the vertical bar within each random effect is allowed to vary by whatever factor is given to the right. Moreover, when multiple terms are listed to the left of the same vertical bar, `lme4` will estimate the correlations between all of these terms. For example, when fitting the random effect '(1 + trial|participant)', `lme4` actually estimates *three* parameters—namely, by-participant varying intercepts, by-participant varying slopes, and the correlation between these two terms.

Using this notation blindly can result in a combinatorial explosion! Let's demonstrate this with an example, a case where a researcher wants to fit by-participant varying slopes for two effects, A and B, as well as by-participant varying slopes in the interactions of these two effects (i.e., it could be that the interaction is expressed differently across participants). This random effects structure embodies the view that participants can vary in both A and B, and the interaction can also vary by participants. The following random effects structure looks rather concise:

```
(A * B|participant)
```

Table 14.1. How to specify different kinds of random effects structures in `lme4` syntax; x is a generic placeholder for any fixed effects predictor (such as 'trial order', 'frequency', 'condition')

lme4 (short)	*lme4 syntax (explicit)*	*Meaning*
x + (1|ppt)	1 + x + (1|ppt)	by-participant varying intercepts
x + (x|ppt)	1 + x + (1+x|ppt)	by-participant varying intercepts and varying x slopes, *with* slope/intercept correlation
x + (x||ppt)	1 + x + (1|ppt) + (0 + x|ppt)	by-participant varying intercepts and varying x slopes, *without* slope/intercept correlation

However, even though the notation looks deceivingly concise, it actually estimates 10 different parameters:

- by-participant varying intercepts ('1')
- by-participant varying A slopes
- by-participant varying B slopes
- by-participant varying A:B interaction slopes
- correlation of varying intercepts and varying A slopes
- correlation of varying intercepts and varying B slopes
- correlation of varying intercepts and varying A:B interaction slopes
- correlation of varying A slopes and varying B slopes
- correlation of varying A slopes and varying A:B interaction slopes
- correlation of varying B slopes and varying A:B interaction slopes

The more explicit notation '(1 + A + B + A:B|participant)' would make some of this clearer, if you remind yourself that correlation terms are estimated for everything to the left of the bar. To suppress these correlation terms, you would have to use the following notation:

```
(1|participant) + (0 + A|participant) +
   (0 + B|participant) + (0 + A:B|participant)
```

The '0' in this notation stands for 'do not fit an intercept'. The fact that the varying intercepts and varying slopes occur in separate brackets is a way of noting that the two should be estimated separately, thus excluding their correlation.

14.8. Reasoning About Your Mixed Model: The Importance of Varying Slopes

A question that comes up many times is: for which variables should one fit varying slopes? You have already seen in Figure 14.1 (the example of Yasmeen, Logan, and Dan) that omitting important varying slopes can lead to gross misrepresentations of the data of particular participants. Within the context of the overall model, this also means that the model doesn't 'know' about the variation between participants, because it wasn't asked to estimate it. This will affect any inferences you want to make based on these estimates. In particular, simulation studies have shown that mixed models that fail to incorporate important varying slope terms can be grossly anti-conservative, having Type I error rates much in excess of the accepted 0.05 threshold (Schielzeth & Forstmeier, 2009).

Linguistics has an interesting history with respect to mixed models. After the publication of Baayen et al. (2008) and Baayen (2008), mixed models took the linguistic community by storm. However, many researchers defaulted to varying intercepts models, not even considering models with varying slopes—even though it was already known in other fields that this was a problematic practice (Schielzeth & Forstmeier, 2009). Barr et al. (2013) made the linguistic community aware of issues that arise from ignoring important varying slopes. However, as a result of their paper, the pendulum swung the other way, and linguists fitted overly complex random effects structures. In particular, their paper used the slogan 'keep it maximal' to argue that researchers should 'maximize' their random effects structures by adding varying slope terms for all critical variables in a study. However, the recommendations by Barr et al. (2013) were often mistaken to suggest that one should maximize blindly and fit varying slopes for every single fixed effect in one's study.

The 'keep it maximal' credo has received some backlash from Matuschek, Kliegl, Vasishth, Baayen, and Bates (2017). As a blind rule to follow, the 'keep it maximal' credo often leads to estimation difficulties, which will be discussed in Chapter 15. Moreover, maximal models may also have lower statistical power (see also Seedorff, Oleson, & McMurray, 2019). The deeper problem, however, is that linguists want simple recipes. Before Barr et al. (2013), researchers often defaulted to varying intercepts models. After Barr et al. (2013), researchers defaulted to 'maximal' models. However, when it comes to statistical modeling, there are no strict recipes that can be followed in all circumstances. Instead, you need to reflect on what the most appropriate model is, given the data and given your theory (see Chapter 16).

This is a good opportunity to walk you through an example of how you can reason about fitting a mixed model. Imagine a researcher is interested in testing whether a simple training manipulation has an effect on reaction times. The condition in this case has two levels: pre-test and post-test (after training). In addition, there are different items (say, words) that are repeated across participants. Two different items are used for the pre-test, and two different items are used for the post-test. The following provides a glance at this dataset.

```
   participant   age item  condition      RT
   <chr>        <dbl> <chr> <chr>       <dbl>
1  P1              32 item1 pre           655
2  P1              32 item2 pre           577
3  P1              32 item3 post          615
4  P1              32 item4 post          625
5  P2              28 item1 pre           616
6  P2              28 item2 pre           596
7  P2              28 item3 post          660
8  P2              28 item4 post          596
```

The researcher then goes on to fit the following mixed effects model (using `lme4` syntax):

```
RT ~ condition + age +
     (1 + condition|participant) +
     (1|item)
```

This model estimates fixed effects of `condition` and `age`, as well as by-participant varying intercepts and by-participant varying condition slopes, including the correlation of these terms. In addition, notice that there are varying intercepts for items in this model, but not varying slopes.

Before determining whether you *should* have varying slopes for participants, you need to determine whether you *can* fit them, for which you have to ask yourself the following question: Does the relevant variable vary within individual? In the present example, it is the case that participants have been exposed to *both* conditions. The fact that there is variation within individuals for the relevant variable (pre-test and post-test scores are both attested) makes it possible to fit varying slopes. The next question is whether these should be fitted.

It seems quite unreasonable to assume that all participants in this study benefit exactly the same way from the training. After all, people are different from each other in all sorts of ways, and experience shows that some people benefit more from certain training regimes than others. Thus, unless you really want to assume that there are no by-participant differences in the training effect, fitting a varying slopes model seems to be the preferred option. Notice that I'm not telling you to simply 'maximize' the random effects structure. Instead, you are adding the by-participant varying slopes because you have thought about the nature of your condition variable and concluded that participants may differ in their condition effects. Thus, your model specification was guided by your knowledge about the phenomenon at hand.

Next, let's go on to the age fixed effect. Does age vary within individual? No, in this case, it doesn't. In the dataset shown above, each participant is of only one age, which means that no varying slopes need to be considered. Notice that whether varying slopes for age make sense depends on the structure of your experiment. For example, if you conducted a longitudinal study where you observed the same individuals across time, then age *does* vary within individuals, which means that it would be possible to estimate by-participant varying slopes of age.

Next, let's move on to the item effect. First of all, in this experiment, items are repeated across individuals, which introduces dependencies. Something unique about a particular item that is unpredictable from the perspective of the experiment may influence all responses to that item, even if they come from separate participants. This warrants including varying intercepts for item. What about by-item varying slopes? As was mentioned above, different items were used for the pre- and post-test conditions in this hypothetical study. As a result of this, condition does not vary within item (this is what's called a 'between-items design' rather than a 'within-items design') and no by-item varying slopes need to be considered for the condition effect.

Finally, let us observe the fact that the age factor *does* vary within item. That is, the same item is combined with different age values across the various participants in this study. So, it is certainly possible to fit by-item varying slopes '(1+age|item)'. However, not everything that can be estimated has to be estimated. Ask yourself a few questions. Are there any compelling reasons to expect by-item varying age effects? Is this theoretically motivated? The answers depends on the nature of the phenomenon that you investigate. In the case of the model formula above, the researcher decided not to estimate by-age varying item effects. Crucially, such a decision should be made in advance, ideally before starting to investigate the data. Chapter 16 will talk more about the ethical issues that arise in the context of specifying your models.

The point of this discussion is that you shouldn't default to either varying intercepts models, and neither should you default to having varying slopes for all variables in your study. There's no substitute for exercising your researcher judgment and thinking about your model. Each term in your model, including the random effects, should be theoretically justified.

14.9. Chapter Conclusions

In this chapter, you've learned about an incredibly powerful new tool: mixed models. These models allow modeling dependent data structures. The 'workhorse' for accounting for dependencies between data points are the random effects. I have guided you through various constellations of these random effects, as well as through the syntax of the most commonly used R package for mixed modeling, lme4. This syntax takes some time to get used to, and the exercises in the next chapter will help you in this respect.

The chapter concluded with a discussion of the importance of varying slopes: Type I error rates increase drastically when important varying slopes are missed, which has severe effects on the inferences you base on your models. Because of that, varying slope terms need to be considered. However, there's no simple rule as to which varying slopes should be included; you have to exercise scientific judgment and critically reflect on what you think is the most appropriate model for a particular analysis, given your knowledge of the phenomenon at hand.

15 Mixed Models 2

Extended Example, Significance Testing, Convergence Issues

15.1. Introduction

The last chapter provided a conceptual introduction to mixed models. This chapter deals with implementation. I need to warn you: the lead-up to this chapter will be rather tedious, as you will generate a reasonably complex dataset yourself, which requires a lot of R programming. However, the conceptual pay-off of working with generated data is that it allows you to see more clearly how different sources of variation are captured by a mixed model. Moreover, being able to create data like this gives you a 'ground truth' to compare a mixed model to, a luxury that you usually don't have.[1]

15.2. Simulating Vowel Durations for a Mixed Model Analysis

In this section, you are going to simulate data that will then be analyzed with a mixed model. By creating random data with exact specifications (such as, 'participants differ from each other with a standard deviation of 20ms'), you are able to assess the extent to which a mixed model can retrieve this information.

Let me specify the desired structure of the data. You are going to simulate a hypothetical experiment with six participants, each of which responds to 20 data points.[2] These data points are the same 20 items repeated across participants. You will investigate the effect of word frequency on vowel durations, where you expect more frequent words to have shorter vowel durations.

Occasionally, we will use the `tidyverse` package in this chapter, so please load this package again in case you have started a new R session:

---

1 Being able to generate data yourself is also a stepping stone towards performing your own Type I and statistical power simulations (see Kirby & Sonderegger, 2018; Brysbaert & Stevens, 2018).
2 A study with six participants is clearly not a high-powered study. As a general rule of thumb, you should never base any inferential statistics on just six participants. The whole concept of making inferences to populations is a very flimsy one with a sample size that low, no matter whether you obtain a significant result or not. However, for the sake of this example, it is useful to keep the numbers low, because it makes it easier to discuss specific cases.

```
library(tidyverse)
```

So that your computer has the same random numbers, let us begin by setting the seed value to a nice number.

```
# Set seed value to make your examples match the book:

set.seed(666)
```

It is important that, from now on, you adhere to the strict sequence of the commands in this chapter. If you execute one of the random number generation functions more than once, you will be 'out of sequence' and get different numbers from those reported in the book (not a big deal, but it may make it more difficult for you to see whether you have implemented everything correctly). If you get stuck, simply re-execute everything up to that point, which should be easy if everything is in the same script.

Let's start by generating participant and item identifiers. For this, the gl() function comes in handy. The name of this function comes from 'generate levels'; it generates factors for a specified number of levels. The first argument of this function specifies the number of levels; the second argument, how many times each level is repeated. The following code generates participant identifiers for six participants, with each level repeated 20 times (for the number of items in the study).

```
ppt_ids <- gl(6, 20)

# Check:

ppt_ids
```

```
  [1] 1 1 1 1 1 1 1 1 1 1 1 1 1 1 1 1 1 1 1 1 2 2 2 2 2 2
 [27] 2 2 2 2 2 2 2 2 2 2 2 2 2 2 3 3 3 3 3 3 3 3 3 3 3 3
 [53] 3 3 3 3 3 3 3 3 4 4 4 4 4 4 4 4 4 4 4 4 4 4 4 4 4 4
 [79] 4 4 5 5 5 5 5 5 5 5 5 5 5 5 5 5 5 5 5 5 5 5 6 6 6 6
[105] 6 6 6 6 6 6 6 6 6 6 6 6 6 6 6 6
Levels: 1 2 3 4 5 6
```

Next, use gl() to create 20 unique item identifiers:

```
# Create 20 item identifiers:

it_ids <- gl(20, 1)

# Check:

it_ids
```

```
 [1] 1  2  3  4 5 6 7 8 9 10 11 12 13 14 15 16
[17] 17 18 19 20
20 Levels: 1 2 3 4 5 6 7 8 9 10 11 12 13 14 ... 2
```

Since each participant responded to all of these items, and there are six participants, this vector needs to be repeated six times. The 'repeat' function rep () was explained in Chapter 8.3.

```
it_ids <- rep(it_ids, 6)

it_ids

 [1]  1  2  3  4  5  6  7  8  9 10 11 12 13 14 15 16 17
[18] 18 19 20  1  2  3  4  5  6  7  8  9 10 11 12 13 14
[35] 15 16 17 18 19 20  1  2  3  4  5  6  7  8  9 10 11
[52] 12 13 14 15 16 17 18 19 20  1  2  3  4  5  6  7  8
[69]  9 10 11 12 13 14 15 16 17 18 19 20  1  2  3  4  5
[86]  6  7  8  9 10 11 12 13 14 15 16 17 18 19 20  1  2
[103]  3  4  5  6  7  8  9 10 11 12 13 14 15 16 17 18 19
[120] 20
20 Levels: 1 2 3 4 5 6 7 8 9 10 11 12 13 14 15 16 ... 20
```

Each participant responds to the items 1 to 20. Let's double-check that the participant and item identifier vectors have the same length.

```
length(ppt_ids)  # 6 participants, each responds to 20 items

[1] 120
```

```
length(it_ids)  # 20 items, each for 6 participants

[1] 120
```

Next, let's create the word frequency predictor. Each word (item) should have its own log frequency value. The rexp() function generates 'random exponentially distributed numbers'. The multiplication by five in the following command is just to make the numbers look more like actual log frequency values (not an important step).

```
# 20 random numbers, rounded:

logfreqs <- round(rexp(20) * 5, 2)

logfreqs

 [1]  2.74  9.82  0.70  0.05  2.76 17.57 13.06
 [8]  7.62 10.38  3.11 14.71  1.62  0.13  3.06
[15]  3.80 15.70  1.12  0.53  1.38  3.05
```

Next, the logfreqs vector needs to be repeated six times, since there are six participants.

```
# Repeat frequency predictor 6 times:

logfreqs <- rep(logfreqs, 6)
```

```
# Check length:

length(logfreqs)
```

```
[1] 120
```

```
# Check content:

logfreqs
```

```
  [1]   2.74   9.82   0.70   0.05   2.76  17.57  13.06   7.62  10.38
 [10]   3.11  14.71   1.62   0.13   3.06   3.80  15.70   1.12   0.53
 [19]   1.38   3.05   2.74   9.82   0.70   0.05   2.76  17.57  13.06
 [28]   7.62  10.38   3.11  14.71   1.62   0.13   3.06   3.80  15.70
 [37]   1.12   0.53   1.38   3.05   2.74   9.82   0.70   0.05   2.76
 [46]  17.57  13.06   7.62  10.38   3.11  14.71   1.62   0.13   3.06
 [55]   3.80  15.70   1.12   0.53   1.38   3.05   2.74   9.82   0.70
 [64]   0.05   2.76  17.57  13.06   7.62  10.38   3.11  14.71   1.62
 [73]   0.13   3.06   3.80  15.70   1.12   0.53   1.38   3.05   2.74
 [82]   9.82   0.70   0.05   2.76  17.57  13.06   7.62  10.38   3.11
 [91]  14.71   1.62   0.13   3.06   3.80  15.70   1.12   0.53   1.38
[100]   3.05   2.74   9.82   0.70   0.05   2.76  17.57  13.06   7.62
[109]  10.38   3.11  14.71   1.62   0.13   3.06   3.80  15.70   1.12
[118]   0.53   1.38   3.05
```

Let us now put all three vectors (participant identifiers, item identifiers, frequencies) into a tibble, naming the three columns participant, item, and freq.

```
# Put predictors together:

xdata <- tibble(ppt = ppt_ids, item = it_ids,
                freq = logfreqs)

xdata
```

```
# A tibble: 120 x 3
    ppt    item    freq
  <fct>   <fct>   <dbl>
 1 1       1       2.74
 2 1       2       9.82
 3 1       3       0.7
 4 1       4       0.05
 5 1       5       2.76
 6 1       6      17.6
 7 1       7      13.1
 8 1       8       7.62
 9 1       9      10.4
10 1      10       3.11
# ... with 110 more rows
```

This tibble contains all the predictors that are of interest to this analysis. Now you need a response, and this is where things become interesting for thinking about how mixed models work. Let's start by specifying a column containing the intercept 300, which seems like a realistic number for vowel duration. Let's add this number to a column called 'int'.

```
xdata$int <- 300
```

It would be unreasonable to assume that each participant has the same intercept, so let's generate six deviation scores (one for each participant) that shift the grand intercept up or down. Let's assume that these deviation scores are normally distributed with a standard deviation of 40 .

```
# Generate varying intercepts for participants:

ppt_ints <- rnorm(6, sd = 40)

# Check:

ppt_ints
```

```
[1]  0.8921713 -0.5763717 53.3142134 5.9370717 3.1084019
[6] 85.1702223
```

The sixth participant, for example, will have intercepts that are above the overall intercept by about +85*ms*.

Before adding these varying intercepts to the tibble, you have to consider the fact that the first 20 rows in the tibble are from participant 1, the next 20 rows are from participant 2, and so on. You need to make sure that the first deviation score is repeated 20 times (for participant 1), then the second score is repeated 20 times (for participant 2), and so on. The each argument of the rep() function is used to specify that *each* number should be repeated 20 times. This ensures that the varying intercepts match up with the participant identifiers.

```
xdata$ppt_ints <- rep(ppt_ints, each = 20)

xdata
```

```
# A tibble: 120 x 5
   ppt   item   freq   int ppt_ints
   <fct> <fct> <dbl> <dbl>    <dbl>
1 1      1      2.74   300    0.892
2 1      2      9.82   300    0.892
3 1      3      0.7    300    0.892
4 1      4      0.05   300    0.892
5 1      5      2.76   300    0.892
6 1      6     17.6    300    0.892
7 1      7     13.1    300    0.892
```

```
 8 1     8       7.62    300     0.892
 9 1     9      10.4     300     0.892
10 1    10       3.11    300     0.892
# ... with 110 more rows
```

Next, you need to create varying intercepts for items, for which another 20 random numbers are needed. Let's say that the item standard deviation is 20ms, deliberately set to a smaller number than the variation across participants.[3]

```
# Generate 20 varying intercepts for items:

item_ints <- rnorm(20, sd = 20)

# Check content:

item_ints
```

```
 [1]   12.779092    5.386949   45.979387  -27.492918  13.247343
 [6]    9.670326   24.645837  -35.552503   17.710548   7.792605
[11]  -12.492652   26.928580   27.854384   -8.894240  22.986354
[16]   16.369922  -33.103585   27.384734   15.164802  20.072252
```

As there are six participants, this vector of item intercepts needs to be repeated six times.

```
# Repeat item intercepts for six participants:

item_ints <- rep(item_ints, times = 6)

# Check that length matches:

length(item_ints)
```

```
[1] 120
```

```
# Add to tibble:

xdata$item_ints <- item_ints
```

The world isn't perfect, and there's always residual variation on top of participant and item variation. So, let's also add trial-by-trial noise, which you can do by generating 120 random numbers, one for each data point:

```
xdata$error <- rnorm(120, sd = 20)
```

3 In my experience it is almost always the case that when you fit a mixed model to real experimental data, participants are more different from each other than items, which results in larger standard deviations for participant random effects than for item random effects.

Up to this point, we have only dealt with random variation. What about the actual frequency effect? Let's assume the slope is −5, which is embodying the assumption that, for each increase in log frequency by 1, vowel durations decrease by 5ms.

```
xdata$effect <- (-5) * xdata$freq

# Check:

xdata %>% head(4)
```

```
# A tibble: 4 x 8
    ppt   item   freq    int ppt_ints item_ints error effect
  <fct> <fct>  <dbl>  <dbl>    <dbl>     <dbl> <dbl>  <dbl>
1 1     1       2.74    300    0.892      12.8  13.9 -13.7
2 1     2       9.82    300    0.892       5.39 -19.1 -49.1
3 1     3       0.7     300    0.892      46.0  22.8  -3.5
4 1     4       0.05    300    0.892     -27.5 -28.3  -0.25
```

Let's behold what you have created. From left to right: participant identifiers (ppt), item identifiers (item), frequency values (freq), an overall intercept (int), participant-varying intercepts (ppt_ints), item-varying intercepts (item_ints), trial-by-trial error (error), and the actual effect (effect). The information in columns int to effect is everything that's needed to create the response variable:

```
xdata <- mutate(xdata,
                dur = int + ppt_ints + item_ints +
                error + effect)
```

Thus, the dur response compresses many sources of variation into one numerical variable. As researchers in this hypothetical study, you would only have the dur column available to you. You don't have the luxury of knowing the true by-participant and by-item variation. The mixed model's task then is to identify the different sources of variation.

Let's dwell on this for a bit longer. I want you to realize that the information in the dur column combines a 'random' part (ppt_ints, item_ints, error), as well as a 'systematic' part (int, effect). The latter is what you're used to from linear regression. The only new thing is that there are now participant- and item-specific variations in the mix. It is the mixed model's task to decompress this information and estimate the individual variance components.

To make this dataset look more like an actual dataset from a real study, let us get rid of those columns that were used to generate the data. By using the minus sign, the select() function can also be used to exclude the consecutive columns from int to effect.[4] Let's save this into a new tibble called xreal.

---

4 If you get an error message when running the select() function call, chances are that you still have the MASS library from Chapter 13 loaded. In that case, you can use: dplyr::select() instead to resolve the naming conflict.

```
xreal <- select(xdata, -(int:effect))

xreal
```

```
# A tibble: 120 x 4
     ppt   item   freq   dur
   <fct> <fct>  <dbl> <dbl>
 1 1      1       2.74  314.
 2 1      2       9.82  238.
 3 1      3       0.7   366.
 4 1      4       0.05  245.
 5 1      5       2.76  304.
 6 1      6      17.6   198.
 7 1      7      13.1   258.
 8 1      8       7.62  241.
 9 1      9      10.4   260.
10 1     10       3.11  298.
# ... with 110 more row
```

This looks like the data that you would get as a researcher who studies how word frequency affects vowel durations. I want you to realize the complexity of what's going on 'behind the scenes' in this case, as I think this is the true conceptual pay-off of having generated the data yourself. You, as the researcher in this case, would've only measured what's in the dur column. But something as simple as measuring durations is affected by a whole lot of different forces. In the next section, you will learn how to use mixed models to help you tease apart what's going on.

In case you're having trouble with the long sequence of commands up to this point, here is all the code that is needed to recreate the data-generation process. If all of the following code is executed sequentially, you will get the same numbers:

```
set.seed(666)
ppt_ids <- gl(6, 20)
it_ids <- gl(20, 1)
it_ids <- rep(it_ids, 6)
logfreqs <- round(rexp(20) * 5, 2)
logfreqs <- rep(logfreqs, 6)
xdata <- tibble(ppt = ppt_ids, item = it_ids, freq = log-
freqs)
xdata$int <- 300
ppt_ints <- rnorm(6, sd = 40)
xdata$ppt_ints <- rep(ppt_ints, each = 20)
item_ints <- rnorm(20, sd = 20)
item_ints <- rep(item_ints, times = 6)
xdata$item_ints <- item_ints
xdata$error <- rnorm(120, sd = 20)
xdata$effect <- -5 * xdata$freq
```

```
xdata <- mutate(xdata,
                dur = int + ppt_ints + item_ints +
                error + effect)
xreal <- select(xdata, -(int:effect))
```

It makes sense to annotate the different steps with comments in your script. Also, notice that I omitted those commands that merely check the results, such as `length(it_ids)`. Such commands are best for quick interactive checks in the console—they don't have to be in the script, since they don't produce any lasting changes. Just in case you got lost somewhere in this long sequence of commands, the file 'frequency_vowel_duration.csv' contains the `xreal` tibble.

15.3. Analyzing the Simulated Vowel Durations with Mixed Models

In this section, you'll analyze the duration data with mixed models using the `lme4` package.[5] Let's begin by loading the package into your current R session.

```
library(lme4)
```

The `lmer()` function is used for specifying mixed models. Let's think about what the model formula should look like. The response is `dur` (vowel durations), and there's only one fixed effect, `freq`. Thus, you know that one part of the formula will be '`dur ~ freq`', duration modeled as a function of the fixed effect frequency.

Let's halt for a moment and think about why frequency should be a fixed effect. The most obvious reason is that frequency is a continuous variable, and only categorical variables can be random effects. An additional reason is that you expect frequency to have a systematic influence on durations, one that you expect to find again if you were to collect a new sample of participants and items (barring some sampling error). Finally, you actually want to measure a coefficient for frequency (a slope), rather than a standard deviation, as would be the case for a random effect.

Participant and item, on the other hand, are more appropriate as random effects. Before you conducted the study, you didn't know how specific participants and items would behave—from the perspective of the experiment, their influence is unpredictable, or 'random'. Rather than estimating separate parameters for specific participants and items, this will estimate the variation around participants and items. Perhaps it helps you to remember that you used the `rnorm()` function to generate the varying intercepts for participants and items.

The following is a linear mixed effects model with a frequency fixed effect and varying intercepts for participants and items. To remind you of the notation '`(1|ppt)`' and '`(1|item)`': what is to the left of the vertical bar (in this case, the intercept represented by '1') varies by the thing to the right of the vertical bar.

---

5 As the lme4 package receives constant updates, it may be the case that the exact results regarding convergence issues discussed in this section may change.

```
xmdl <- lmer(dur ~ freq + (1|ppt) + (1|item),
              data = xreal, REML = FALSE)
```

What does the `REML = FALSE` mean? 'REML' stands for restricted maximum likelihood. By setting this argument to `FALSE`, the model will use what's called 'maximum likelihood' for estimation. This is a technical detail that is important when you want to use mixed models for testing the significance of fixed effects using likelihood comparisons (see Pinheiro & Bates, 2000), as will be done later in this chapter.

Now that you have the model saved, it's time to inspect it.

```
summary(xmdl)
```

```
Linear mixed model fit by maximum likelihood ['lmerMod']
Formula: dur ~ freq + (1 | ppt) + (1 | item)
   Data: xreal

     AIC      BIC    logLik deviance  df.resid
  1105.1   1119.0    -547.5   1095.1       115

Scaled residuals:
     Min       1Q    Median       3Q      Max
-2.09700 -0.60946  0.06483  0.60761  2.39754

Random effects:
 Groups   Name        Variance Std.Dev.
 item     (Intercept)  589.2    24.27
 ppt      (Intercept) 1296.6    36.01
 Residual              284.0    16.85
Number of obs: 120, groups:  item, 20; ppt, 6

Fixed effects:
            Estimate Std. Error t value
(Intercept)  337.973     16.735  20.196
freq          -5.460      1.004  -5.438

Correlation of Fixed Effects:
     (Intr)
freq -0.339
```

The output reminds you that you fitted a linear mixed effects model with maximum likelihood (`REML = FALSE`), and it also restates the model formula. Following this, there are some general measures of model fit (Aikaike's Information Criterion, Bayesian Information Criterion, log likelihood, and the degrees of freedom). At this stage, these numbers do not concern us much. However, it's worth mentioning that the 'df.resid' represents the residual degrees of freedom, which are the number of data points minus the number of estimated parameters. This turns out to be 115, as the model was fitted to 120 data points and estimated five parameters. In this case, each estimated parameter corresponds to a row in the random effects output (`item`, `ppt`, and `Residual`), as well as to a row in the fixed effects output (`(Intercept)` and `freq`).

Let's first focus our eyes on the fixed effects estimates. First of all, the intercept in the fixed effects component is about 340ms, which means that, for a log frequency of 0, this model predicts vowels to be 340ms long. The frequency slope is negative (–5.46) and very close to what was specified during data generation. The usual interpretation to slopes applies in this case: for each increment in frequency by one unit, vowels become about 5ms shorter.

Let's put the fixed effects coefficients into a familiar format:

$$duration = 338 + (-5) \times frequency \qquad \text{(E15.1)}$$

Just as you did in other chapters, I invite you to plug some numbers into this equation to see what it predicts. For example, for a word frequency of 10, the equation becomes *duration* = 338*ms* − 50*ms*, which is 288ms. Thus, for a word frequency of 10, vowels are predicted to be 288ms long.

Let's move our eyes over to the random effects components. Notice that the estimate of the by-participant variation in intercepts (36.01) is larger than the by-item variation in intercepts (24.287) and the residual variation (16.85), just as was specified during data generation

15.4. Extracting Information out of lme4 Objects

It's a good idea to learn about a few subsidiary functions that can be used to extract information from fitted mixed model objects. First, the fixef() function spits out the estimated fixed effects coefficients.

```
fixef(xmdl)
```

```
(Intercept)         freq
337.973044     -5.460115
```

```
fixef(xmdl)[2] # extract slope
```

```
freq
-5.460115
```

The following command retrieves the coefficient table from the summary output.[6] Unfortunately, the broom package for tidy regression output doesn't work so well for mixed models.

```
summary(xmdl)$coefficients
```

```
              Estimate Std. Error    t value
(Intercept) 337.973044  16.734965  20.195624
freq         -5.460115   1.004059  -5.438042
```

6 In case you didn't know that the coefficients information in the summary output can be retrieved by $coefficients, the str() function helps: str(summary(xmdl)). The str() function shows the general structure of any R object you apply it to.

What happens if you apply `coef()` to the model object?

```
coef(xmdl)
```

```
$item
   (Intercept)       freq
1     352.2840  -5.460115
2     325.2502  -5.460115
3     370.0176  -5.460115
4     302.6063  -5.460115
5     349.9889  -5.460115
6     338.9433  -5.460115
7     362.7144  -5.460115
8     295.8086  -5.460115
9     333.0941  -5.460115
10    331.7932  -5.460115
11    324.7939  -5.460115
12    350.1699  -5.460115
13    353.0408  -5.460115
14    311.9676  -5.460115
15    353.9778  -5.460115
16    353.9778  -5.460115
17    289.0330  -5.460115
18    362.3463  -5.460115
19    338.1415  -5.460115
20    359.6313  -5.460115


$ppt
   (Intercept)       freq
1     315.3049  -5.460115
2     301.6252  -5.460115
3     363.4343  -5.460115
4     318.0924  -5.460115
5     324.4672  -5.460115
6     404.9142  -5.460115

attr(,"class")
[1] "coef.mer"
```

The function spits out the random effects estimates for specific participants and items. Notice that, when applied to mixed models, the `coef()` function behaves differently than in the context of `lm()`. For linear models, the function returns the fixed effects coefficients; for linear mixed effects models, it returns the random effects estimates! The `coef()` output is a list, with one list element for every grouping factor. Each list element contains a data frame, which you can retrieve via the dollar sign '$'. The following command extracts the data frame with the random effects estimates for the individual participants.

```
coef(xmdl)$ppt
```

```
   (Intercept)      freq
1     315.3049  -5.460115
2     301.6252  -5.460115
3     363.4343  -5.460115
4     318.0924  -5.460115
5     324.4672  -5.460115
6     404.9142  -5.460115
```

The slowest speaker is participant 6 (a high intercept of around 405ms); the fastest speaker is participant 2 (around 302ms). Whereas there are different numbers for each participant, the numbers in the frequency column are all the same. This is because this is a varying intercept, but not a varying slope model. The frequency effect does not vary across participants; it's assumed to be *fixed*. By-participant variation in frequency slopes is not estimated, because the model hasn't been specified to look for random slopes.

The ranef() function is similar to coef(), but it spits out the deviations from the intercept, rather than the actual intercept estimates. You can then see, for example, that participant 6 is estimated to have an intercept that is +67ms above the grand intercept, making this participant a slow speaker.

```
ranef(xmdl)$ppt
```

```
   (Intercept)
1    -22.66813
2    -36.34780
3     25.46129
4    -19.88066
5    -13.50587
6     66.94117
```

15.5. Messing up the Model

Let's play a bit more with different model specifications. I don't recommend 'playing' with different models in any actual data analysis (see Chapter 16), since you should usually have decided about your model in advance of conducting a data analysis (based on theoretical reasoning). However, for demonstration purposes only, it'll be useful to use to see what goes wrong if you 'mess up' the model.

For example, what if you were to drop the term for estimated by-item varying intercepts? Let's call the resulting model xmdl_bad.

```
# Fit model that drops important item-varying intercepts:

xmdl_bad <- lmer(dur ~ freq + (1|ppt),
                 data = xreal, REML = FALSE)

# Check output:

summary(xmdl_bad)
```

```
Linear mixed model fit by maximum likelihood ['lmerMod']
Formula: dur ~ freq + (1 | ppt)
   Data: xreal


     AIC      BIC    logLik deviance df.resid
   1182.1   1193.2   -587.0   1174.1      116


Scaled residuals:
    Min      1Q   Median      3Q      Max
 -2.6270 -0.6397  0.1089  0.7462  1.9778


Random effects:
 Groups   Name         Variance Std.Dev.
 ppt      (Intercept)  1240.8   35.23
 Residual              877.6    29.62
Number of obs: 120, groups: ppt, 6


Fixed effects:
             Estimate Std. Error t value
(Intercept) 337.9730    14.8829   22.71
freq         -5.4601     0.4813  -11.34


Correlation of Fixed Effects:
     (Intr)
freq -0.183
```

First, notice that you are now estimating one parameter less, as reflected in the 'df.resid' having changed from 115 to 116. As there are 120 data points, this shows that there are four parameters being estimated in this case ($120 - 4 = 116$). This is also shown by the fact that there is one row less in the random effects component of the output.

Notice, furthermore, that the residual standard deviation has almost doubled compared to the previous model, increasing from ~17ms to ~30ms. Finally, notice that the standard error of the frequency coefficient has halved (from $SE = 1.0$ to $SE = 0.48$). Taken at face value, this would suggest that your accuracy in estimating the slope has increased by dropping the item random effects. However, in fact, this is just your model becoming more anti-conservative, because it doesn't 'know' about important sources of variation in the data (as discussed in Chapter 14).

Next, what happens if you add by-participant-varying slopes to the model? Let us add by-participant varying slopes to a model that we call xmdl_slope.[7] Below, you look at only the random effects output, which can be extracted by adding $varcor to the summary() output.

---

7 Items cannot vary in the slope of frequency, because each item has only one frequency value. So it makes no sense to estimate by-item varying slopes in this particular case.

```
# Fit model with frequency slope:

xmdl_slope <- lmer(dur ~ freq + (1 + freq|ppt) + (1|item),
                   data = xreal, REML = FALSE)

boundary (singular) fit: see ?isSingular
```

summary(xmdl_slope)$varcor

```
Groups    Name         Std.Dev. Corr
item      (Intercept)  24.27168
ppt       (Intercept)  35.25896
          freq          0.12109 1.000
Residual               16.83902
```

The warning message on 'singular fits' suggests that there are some problems with this model. Estimating varying slopes added an additional line to the random effects output. Notice that the estimated standard deviation across varying slopes is very small. Why is this? Well, you didn't specify any varying slopes when the data was constructed! Remember that you added by-participant and by-item varying intercepts, but you didn't make slopes dependent on participants. Nevertheless, this model tries to estimate by-participant variation in frequency slopes. The fact that there is any variance that can be attributed to this at all is a by-product of adding random noise to the data.

There's a hint in this output which indicates that this model fits badly. Notice that the varying slope intercept/correlation is indicated to be *exactly* +1. If this were true, this would mean that higher intercepts always (deterministically) go together with higher frequency slopes (the correlation is indicated to be negative). Perfect correlations never happen in linguistic data. This seems suspicious. It turns out that lme4 sometimes fixes random effect correlation terms to +1.0 or –1.0 if these cannot be estimated (Bates et al., 2015; see also discussion in Vasishth, Nicenboim, Beckman, Li, & Kong, 2018). In other words, exact correlations like this are a warning flag that suggests estimation problems. This also relates to the 'singular fit' message in the output above (see discussion below on convergence issues).

You can also check the random effects estimates of this model, which reveals that, although the frequency slopes are now slightly different from each other, there is very little variation between participants in their respective slopes.

coef(xmdl_slope)$ppt

```
  (Intercept)       freq
1   315.7932  -5.536287
2   302.3495  -5.582457
3   362.9137  -5.374462
4   318.3863  -5.527382
5   324.7295  -5.505597
6   403.6660  -5.234507
```

Let us now specify a model with the de-correlated random effects structure. As was mentioned in the previous chapter, this is achieved by separating the intercept and slope terms into separate brackets. You use the zero character in (0 + freq|ppt) to signal that you're removing the intercept for this term.

```
xmdl_slope_nocorr <- lmer(dur ~ freq +
                          (1|ppt) + (1|item) +
                          (0 + freq|ppt),
                          data = xreal, REML = FALSE)
```

```
boundary (singular) fit: see ?isSingular
```

```
summary(xmdl_slope_nocorr)$varcor
```

```
Groups     Name          Std.Dev.
item       (Intercept)   24.288
ppt        freq           0.000
ppt.1      (Intercept)   36.004
Residual                 16.852
```

The correlation term has disappeared, but in this case, the frequency slope is estimated to have a standard deviation of exactly 0.0, which means that it could not be estimated. This is also apparent when checking the random effects with coef(), where all slopes in the freq column are the same.

```
coef(xmdl_slope_nocorr)$ppt
```

```
    (Intercept)        Freq
1      315.3049   -5.460115
2      301.6253   -5.460115
3      363.4343   -5.460115
4      318.0924   -5.460115
5      324.4672   -5.460115
6      404.9142   -5.460115
```

It is important to dwell on this point a little longer. Yet again, there was no error or warning message; nevertheless lme4 failed to estimate the by-participant varying slopes. If you didn't inspect the random effects structure in detail, this would go unnoticed. You may have reported this as a random slope model, even though in fact random slopes weren't estimated.

15.6. Likelihood Ratio Tests

This section teaches you how to perform likelihood ratio comparisons (otherwise known as 'deviance tests') to perform significance tests with mixed models. A likelihood ratio test compares the likelihood of one model to the likelihood of another model. What's the likelihood? In everyday language, the terms 'probability' and 'likelihood' are often used interchangeably. In statistics, however, these two differ in meaning. The term 'probability' describes the probability of a particular outcome *given a parameter*. For example, how probable it is to observe a count of 3 given a Poisson with $\lambda = 2$. The likelihood describes *the plausibility of particular parameter values, given a set of data.* For example, how likely is the parameter $\lambda = 2$ given an observed count of 3? Mixed models are estimated using 'maximum likelihood estimation', i.e., they find the parameter estimates that are most likely, given a set of observations.

If you wanted to compute the significance of the `freq` fixed effect, you need to compare a model with this effect against a model without it, which is an intercept-only model in this case:

```
xmdl_nofreq <- lmer(dur ~ 1 + (1|ppt) + (1|item),
                    data = xreal, REML = FALSE)
```

The `anova()` function is used to perform model comparisons. The name of this function comes from 'analysis of variance' (see Chapter 11.3), but this function actually performs a likelihood ratio test when applied to mixed models. The only two arguments needed are the two models to be compared:[8]

```
anova(xmdl_nofreq, xmdl, test = 'Chisq')

Data: xreal
Models:
xmdl_nofreq: dur ~ 1 + (1 | ppt) + (1 | item)
xmdl: dur ~ freq + (1 | ppt) + (1 | item)
            Df    AIC    BIC  logLik deviance Chisq Chi Df
xmdl_nofreq  4 1121.0 1132.2 -556.51   1113.0
xmdl         5 1105.1 1119.0 -547.55   1095.1 17.933      1
            Pr(>Chisq)
xmdl_nofreq
xmdl         2.288e-05 ***
---
Signif. codes: 0 '***' 0.001 '**' 0.01 '*' 0.05 '.' 0.1 ' ' 1
```

The p-value is very small ($2.288e-05$), which means the following: under the null hypothesis that the two models are the same, the actually observed difference in likelihood between the two models is surprising. In other words, there is sufficient evidence against the null hypothesis of model equivalence. In this output, the degrees of freedom next to each model indicate the number of estimated parameters, and the degrees of freedom associated with the Chi-Square value is the difference in the number of estimated parameters between the two models. This turns out to be 1 in this case, as one model estimates five parameters and the other model estimates four parameters.

---

8 The `test = 'Chisq'` argument can actually be dropped (it is the default for mixed models), but I prefer to leave it in for clarity. You may wonder what this argument means. This is very technical (so feel free to skip this), but it turns out that −2 times the log likelihood difference between two models is approximately Chi-Square distributed. In Chapter 9, you saw the t-distribution, which is the distribution for the t-test statistic under the null hypothesis. Similarly, there is a Chi-Square distribution, for which the area under the curve also yields a p-value.

The result of this likelihood ratio test can be written up as follows: 'A likelihood ratio test of the model with the frequency effect against the model without the frequency effect revealed a significant difference between models ($\chi^2(1) = 17.93, p < 0.0001$).' Or perhaps, if you stated earlier in the paper that all *p*-values were generated via likelihood ratio tests, you could just say 'There was a significant effect of frequency ($\chi^2(1) = 17.93, p < 0.0001$).' Notice that the degrees of freedom goes inside the brackets after the Chi-Square value.

For many experimental situations, you want to keep the random effects structure constant across comparisons. Sometimes, however, it may be useful to test for the significance of particular random effects. When doing this, f one should use 'restricted maximum likelihood' (REML = TRUE) rather than maximum likelihood (Pinheiro & Bates, 1980). First, let's create two models that only differ in terms of their random effects structure.

```
# Full model:

x_REML <- lmer(dur ~ freq + (1|ppt) + (1|item),
               data = xreal, REML = TRUE)

# Reduced model (no item-varying intercepts):

x_red <- lmer(dur ~ freq + (1|ppt),
              data = xreal, REML = TRUE)
```

Second, let's perform the model comparison. However, since the default of anova() is to fit models with REML = FALSE, the argument refit = FALSE is needed to prevent refitting the models.

```
anova(x_red, x_REML, test = 'Chisq', refit = FALSE)

Data: xreal
Models:
x_red: dur ~ freq + (1 | ppt)
x_REML: dur ~ freq + (1 | ppt) + (1 | item)
        Df    AIC     BIC  logLik deviance   Chisq Chi Df
x_red    4 1174.4 1185.5  -583.2   1166.4
x_REML   5 1095.8 1109.7  -542.9   1085.8  80.608      1
        Pr(>Chisq)
x_red
x_REML   < 2.2e-16 ***
---
Signif. codes: 0 '***' 0.001 '**' 0.01 '*' 0.05 '.' 0.1 ' ' 1
```

The output shows a significant *p*-value, which suggests that dropping the item random effect leads to a significant decrease in likelihood. In other words, there is sufficient evidence against the null hypothesis of model equivalence, or, the random effect is 'significant'.

For complex models with many fixed effects, it may be quite cumbersome to derive *p*-values for each predictor via likelihood ratio tests, as this requires specifying many null models. The `mixed()` function from the `afex` package (Singmann, Bolker, Westfall, & Aust, 2016) performs likelihood ratio tests automatically for all fixed effects when the argument `method = 'LRT'` is specified. The model also automatically sum-codes categorical predictors (see Chapter 7) and warns you of continuous predictors that haven't been centered (see Chapter 5), as this makes the interpretation of interactions difficult (Chapter 8).

```
library(afex)

xmdl_afex <- mixed(dur ~ freq + (1|ppt) + (1|item),
                   data = xreal, method = 'LRT')

Contrasts set to contr.sum for the following variables:
ppt, item
Numerical variables NOT centered on 0: freq
If in interactions, interpretation of lower order (e.g.,
main) effects difficult.
REML argument to lmer() set to FALSE for method = 'PB' or
'LRT'
Fitting 2 (g)lmer() models:
[..]
```

Typing `xmdl_afex` presents a table of the likelihood ratio test results.

```
xmdl_afex

Mixed Model Anova Table (Type 3 tests, LRT-method)
Model: dur ~ freq + (1 | ppt) + (1 | item)
Data: xdata
Df full  model: 5
  Effect df   Chisq p.value
1  freq 1  17.93 *** <.0001
---
Signif. codes: 0 '***' 0.001 '**' 0.01 '*' 0.05 '+' 0.1 ' ' 1
```

The `xmdl_afex` object created by the `mixed()` function contains the full model as well as all 'nested' models that were used for likelihood ratio tests. One can index the full model with `$full_model`, such as in the following command:

```
fixef(xmdl_afex$full_model)

(Intercept)      freq
337.973044   -5.460115
```

When performing model comparisons with likelihood ratio tests, it is a requirement that the comparison models are 'nested', which means that when a reduced model is compared to a full model, the reduced model needs to also be a part of the full model. This is best illustrated with an example. Let's assume the following is your full model (ignoring any random effects for the time being):

```
y ~ A + B
```

This full model can be compared against either one of the following two models:

```
y ~ A
y ~ B
```

This is because the terms in these reduced models are also contained in the full model. However, for reasons not discussed here, you cannot use likelihood ratio tests to compare 'y ~ A' and 'y ~ B', as these models are not nested within each other.

15.7. Remaining Issues

15.7.1. R^2 for Mixed Models

You may have noticed that the linear mixed effects model output does not list R^2-values. You can use the r.squaredGLMM() function from the MuMIn package (Bartoń, 2017) to generate R^2 values for mixed models. The following code applies this function to the full model from the afex model object.

```
library(MuMIn)

r.squaredGLMM(xmdl_afex$full_model)
```

```
      R2m         R2c
0.3043088 0.9089309
```

R2m is the 'marginal R^2' value and characterizes the variance described by the fixed effects. R2c is the 'conditional R^2' value and characterizes the variance described by both fixed and random effects (see Nakagawa & Schielzeth, 2013). It's worth noting that r.squaredGLMM() can be used on generalized linear models that are not mixed models, such as the simple logistic and Poisson regression models we discussed in Chapters 12 and 13.

15.7.2. Predictions from Mixed Models

The following code uses the model discussed earlier to create predictions for a sequence of numbers ranging from 0 to 18 in a step-size of 0.01. The predictions are generated using the equation $y = b_0 + b_1 * freq$.

```
xvals <- seq(0, 18, 0.01)

yvals <- fixef(xmdl_afex$full_model)[1] +
```

```
fixef(xmdl_afex$full_model)[2] * xvals
```

```
head(yvals)
```

```
[1] 337.9730 337.9184 337.8638 337.8092 337.7546 337.7000
```

However, getting confidence intervals for these predictions is more involved and won't be covered in this book. For this, I recommend having a look at the emmeans package (Lenth, 2018). In addition, the mixed models wiki (http://bbolker.github.io/mixedmodels-misc/glmmFAQ.html [accessed October 22, 2018]) contains helpful information, as well as code for generating predictions that can be adapted to specific models.

15.7.3. Convergence Issues

With lme4, you will frequently run into what are called 'convergence issues'. The procedures used to estimate mixed models sometimes do not reach a stable solution; or they reach a stable solution only for a subpart of the parameter space that turned out to be estimable. To understand how this happens, it helps to have a rough mental image of how likelihood estimation works.

Imagine the following scenario: you are being parachuted into an unknown landscape with the task of finding the biggest hill.[9] But there's a catch: you have to find the hill being blindfolded, which means that you cannot simply walk towards the hill. In the absence of sight, one possible approach of finding the hill is to perform a step in a random direction and notice whether your height has increased. If the step has led to an increase in height, stay there and make this your new starting point for the next random step. If the step didn't lead to a noticeable change in height, or perhaps even a reduction, go back one step. This procedure can converge on the hill.

In likelihood estimation, then, the hill is the maximum likelihood estimate, and the blindfolded parachutist following a particular set of instructions is the optimization algorithm that searches for the hill. The same way that the blindfolded parachutist cannot see the hill from a distance, your mixed model doesn't know in advance for which parameter estimates you will obtain the maximum likelihood estimate. So different parameter values are explored by the algorithm to see whether they lead to a noteworthy increase or decrease in likelihood, until a stable solution has been reached.

The more parameters you ask your models to estimate, the more complex the estimation problem becomes. This can be likened to a situation where the blindfolded explorer has to explore an increasingly unwieldy landscape in ever more directions. It could happen that, no matter in which direction the parachutist walks, there is no noteworthy change in height, or there the height changes too drastically for each step.

If convergence isn't reached, lme4 will often spit out a convergence warning. However, as you have already seen in this chapter, there are also estimation problems without any warnings. This is why it is absolutely crucial to investigate the random effects

---

9 I first heard about this analogy in Sarah Depaoli's stats class at UC Merced.

structure—for example, using functions such as `coef()`, and checking that random effect correlation parameters are appropriately estimated (not exactly 0, 1, or –1).

In general, the more complex a model is relative to the size of a dataset, the more difficult it is to estimate models. Convergence issues often arise from trying to fit overly complex models to sparse data; however, they may arise from other issues as well. The problem is that a textbook cannot give you a 'one-size-fits-all' solution to all convergence issues. This is because a convergence issue always stems from the unique combination of the data and the model. It's impossible to give clear recommendations that will apply to all cases.

A useful starting point is the following help page on convergence issues from the `lme4` package.

```
?convergence
```

Some of the recommendations mentioned on that page are worth highlighting here. First, centering and/or scaling continuous variables often facilitates convergence (and sum-coding may achieve similar effects in some cases). Second, you can change the optimizer (the procedure that is used to estimate the model) using the argument `lmerControl`. The help page `?lmerControl` provides useful information on changing the optimizing algorithm. In addition, consider running the `all_fit()` function from the `afex` package on your model. This function refits a model for a whole range of optimization algorithms ("optimizers"), each time assessing convergence. The model will tell you for which optimizers convergence was reached.

However, more fundamentally, obtaining a convergence warning invites you to think about a given model and the structure of your data more deeply. For example, it may be that you have misspecified the model, such as asking the model to estimate varying slopes for a fixed effect that does not actually vary for that grouping factor. There may also be imbalances or sparsity in the dataset worth thinking about. For example, is there perhaps a participant or item with lots of missing values?

As a last resort, you may have to consider simplifying the model. In some cases, the data just is not informative enough to support a specific model (for a linguistic example, see Jaeger et al., 2011). You will often find that random slopes are the cause of many convergence issues. Importantly, you should not drop random slopes lightheartedly, as this is known to increase the Type I error rate (see Barr et al., 2013; Seedorff, Oleson, & McMurray, 2019). An intermediate solution is to explore whether the model converges with a random effects structure for which there is no slope/intercept correlation term. For example, for the factor 'condition', one could fit '(1 | subject) + (0 + condition|participant)' instead of '(1 + condition|subject)'. While some simulation studies show this to be an appropriate intermediate solution (Seedorff et al., 2019), one cannot say that this is a 'safe' strategy in all circumstances, as this depends on whether there are actually slope/intercept correlations in the data.

Another potential solution to convergence issues is to aggregate the data, e.g., computing averages per some grouping factor in the study, such as by-participant averages. While this can be an appropriate solution in some circumstances (see Seedorff et al., 2019), it is a fundamental change of analysis approach that has several important ramifications. Among other things, averaging over trials, one loses the ability to make trial-level predictions, and those sources of variation that are averaged out are simply ignored, even though they may be theoretically important. In addition, this approach generally reduces statistical power.

It turns out that Bayesian mixed models allow dealing with non-convergence in a much more principled fashion, and many estimation problems can be avoided (Eager & Roy, 2017). These models will not be covered in this book, but the reader is advised to consider Vasishth et al. (2018) for a tutorial introduction to Bayesian mixed models (see also Nicenboim & Vasishth, 2016). For a general introduction to Bayesian statistics focusing on conceptual issues, see Dienes (2008). A comprehensive introduction to Bayesian statistics with R is presented in McElreath (2016).

It's best if you can pre-empt convergence issues at the design stage of your study by making sure that there is lots of data. Chapter 10 talked about how important it is to conduct high-powered studies (to avoid Type II errors, Type M errors, and so on). So here is yet another reason for collecting more data: it alleviates convergence issues because, with more informative data, more complex models can be estimated.

15.8. Mixed Logistic Regression: Ugly Selfies

I want to conclude with one example analysis of a real dataset. This example exemplifies convergence issues, as well as how to fit mixed logistic regression models. Everything you have learned about generalized linear mixed effects models in Chapters 12 and 13 carries over to the case of mixed effects models. If you want to fit a Poisson regression or logistic regression, use `glmer()` instead of `lmer()` and specify `family = 'poisson'` or `family = 'binomial'`.

You will be analyzing a small subset of unpublished data that Ruth Page (University of Birmingham) and I collected together. This study investigates 'ugly selfies': people (especially young folks) take pictures in which they portrait themselves as ugly. This phenomenon is theoretically interesting to pragmatics and discourse analysis, as it involves such theoretical constructs as sarcasm and the communication of intimacy. In the experiment we conducted, we were interested whether selfies with different camera angles (either 'from below' or 'level') were perceived as more or less ugly. Let's have a look at the data:

```
selfie <- read_csv('ruth_page_selfies.csv')

selfie
```

```
# A tibble: 1,568 x 3
    Angle      UglyCat    ID
    <chr>      <chr>      <chr>
 1  FromBelow  ugly       ppt_1
 2  FromBelow  not_ugly   ppt_1
 3  Level      not_ugly   ppt_1
 4  Level      not_ugly   ppt_1
```

```
 5 FromBelow not_ugly ppt_1
 6 FromBelow not_ugly ppt_1
 7 Level     not_ugly ppt_1
 8 Level     not_ugly ppt_1
 9 FromBelow not_ugly ppt_1
10 FromBelow not_ugly ppt_1
# ... with 1,558 more rows
```

The column that contains the response is called `UglyCat`, which stands for 'ugly categorical'.[10] The column that contains the predictor is called `Angle`. Participant identifiers are in the column `ID`. Each participant delivered multiple responses, which introduces dependencies. Therefore, a mixed model is needed. Moreover, since the response is categorical, a mixed logistic regression model is needed.

Before fitting a mixed logistic regression model, the `UglyCat` column needs to be converted into a factor (see Chapter 12.6.2).

```
selfie <- mutate(selfie, UglyCat = factor(UglyCat))
```

What should the model formula be? As the main research question is whether ugly judgments are influenced by the camera angle, the formula should contain 'UglyCat ~ Angle'. With respect to the random effects structure, random intercepts are clearly needed, as it is plausible that some participants may have overall more or less 'ugly' judgments. So, at a bare minimum, the mixed model needs to also contain the term (1|ID).

To make a decision about whether random slopes for `Angle` are needed, ask yourself whether it varies within individuals. As each participant responded to both camera angles, the answer to this question is 'yes'. This means that you *can* fit varying slopes for the factor `Angle`. Moreover, as it is unreasonable to assume that all participants are affected by the angle manipulation the same way, it makes sense to fit varying slopes. Thus, this seems like a clear case that calls for including by-participant varying slopes for the condition effect.

```
ugly_mdl <- glmer(UglyCat ~ Angle +
                     (1 + Angle|ID), data = selfie,
                  family = 'binomial')
```

```
Warning messages:
1: In checkConv(attr(opt, "derivs"), opt$par,
ctrl = control$checkConv, :
 Model failed to converge with max|grad| = 0.112061
 (tol = 0.001, component 1)
2: In checkConv(attr(opt, "derivs"), opt$par,
```

---

10 The actual study used a rating scale from 1 to 5. Here, I partitioned the response into a binary categorical variable for pedagogical purposes, so that a mixed logistic regression model can be fitted.

```
ctrl = control$checkConv, :
  Model is nearly unidentifiable: very large eigenvalue
  - Rescale variables?
```

The convergence warning indicates that the model cannot be trusted, and you should not report results from this model in a publication. I used the `all_fit()` function from the `afex` package to find an optimizer that works.

```
all_fit(ugly_mdl)    # abbreviated output:
```

```
bobyqa. : [OK]
Nelder_Mead. : [OK]
optimx.nlminb : [ERROR]
optimx.L-BFGS-B : [ERROR]
nloptwrap.NLOPT_LN_NELDERMEAD : [OK]
nloptwrap.NLOPT_LN_BOBYQA : [OK]
nmkbw. : [OK]
```

The `all_fit()` function indicates that the 'bobyqa' optimizer is one of the optimizers that leads to successful convergence. The following command refits the model with this optimizer.

```
ugly_mdl <- glmer(UglyCat ~ Angle +
                  (1 + Angle|ID), data = selfie,
                  family = 'binomial',
                  control =
                  glmerControl(optimizer = 'bobyqa'))
```

Notice that the convergence warning has gone away. This model can be reported. Let's inspect the model.

```
summary(ugly_mdl)
```

```
Generalized linear mixed model fit by maximum likelihood
  (Laplace Approximation) [glmerMod]
  Family: binomial (logit)
Formula: UglyCat ~ Angle + (1 + Angle | ID)
   Data: selfie
Control: glmerControl(optimizer = "bobyqa")

   AIC      BIC   logLik deviance df.resid
 1712.5   1739.2  -851.2   1702.5     1562

Scaled residuals:
    Min      1Q  Median      3Q     Max
-2.8757 -0.5489  0.1733  0.5502  3.0998
```

```
Random effects:
 Groups Name         Variance Std.Dev. Corr
 ID     (Intercept)  3.61     1.900
        AngleLevel   1.89     1.375    -0.38
Number of obs: 1567, groups:  ID, 98
```

```
Fixed effects:
            Estimate Std. Error z value Pr(>|z|)
(Intercept)   0.9049     0.2199   4.115 3.88e-05 ***
AngleLevel   -1.4834     0.2001  -7.414 1.22e-13 ***
---
Signif. codes:  0 '***' 0.001 '**' 0.01 '*' 0.05 '.' 0.1 ' ' 1
Correlation of Fixed Effects:
           (Intr)
AngleLevel -0.486
```

In the case of mixed logistic regression and mixed Poisson regression, lme4 will spit out a p-value that is based on a 'Wald test'. This test is computationally more efficient than a likelihood ratio test. Wald tests and the likelihood ratio test produce very similar results for very large sample sizes; however, likelihood ratio tests are generally the preferred option for various reasons (e.g., Hauck Jr. & Donner, 1977; Agresti, 2002: 172); among others, the likelihood ratio tests have higher statistical power (Williamson, Lin, Lyles, & Hightower, 2007; Gudicha, Schmittmann, & Vermunt, 2017). So, an alternative way of getting p-values for the Angle effect reported above would be to construct a null model without this predictor (leaving the random effects structure constant), and running a model comparison with the anova() function (see Exercise 15.11.1).[11]

As this is a logistic regression model, the coefficients are log odds (Chapter 12). To interpret the model, it helps to know what is the quantity that is being predicted:

```
levels(selfie$UglyCat)
```

```
[1] "not_ugly" "ugly"
```

Since 'n' precedes 'u' in the alphabet, 'not_ugly' is made the reference level, which means that the log odds are reported in terms of observing an 'ugly' response. As the AngleLevel coefficient is negative (-1.4834), this means that level angles led to a decrease in ugly responses. Chapter 12 discusses how to compute probabilities.

An important take-home message of this example analysis is that a convergence warning should never be ignored (see Matuschek et al., 2017; Seedorff et al., 2019). You cannot report the results of a non-converging model in a published study, as you won't know whether things have been estimated correctly or not.

---

11 It should be noted that the p-values provided by the regression outputs in Chapters 12 and 13 for logistic regression and Poisson regression are also based on Wald tests. In these cases, likelihood ratio tests are also preferred.

15.9. Shrinkage and Individual Differences

To round off this chapter, I want to pre-empt some potential misunderstandings my discussion of random effects may have caused up to this point. So far, we have been concerned with having the right random effects structure in order to make reliable inferences for fixed effects predictors. When investigating the random effects structure, we have focused on making sure that everything is estimated correctly. In my own experience, researchers in the language sciences often focus too much on the fixed effects and, in doing so, they throw away a lot of useful information from their mixed models.

Figure 15.1 shows data from a repeated measures study with multiple responses per participant. You can think of this as a frequency effect, where more frequent words lead to shorter response durations, but participants differ in the extent to which they are affected by word frequency.

Now, imagine that, rather than using a mixed model, you simply performed separate regression analyses for each individual participant. The problem with this approach is that none of the individual regression models knows about what the other models estimate. For example, information in estimating participant A's slope is not used in estimating participant B's slope; that is, the population-level perspective is completely lost when separate models are fitted.

On the other hand, the mixed model takes the information from the entire dataset into account. The varying intercepts and varying slopes are estimated together with the population-level estimates. As a result of this, information about the population is used in estimating individuals. This results in the individual estimates being drawn towards the population-level estimate, a feature of mixed models that is also known as 'shrinkage', visualized in Figure 15.1b. Notice how the intercepts are 'shrunken'

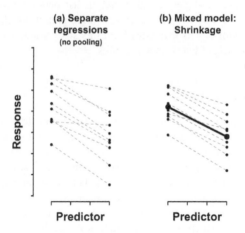

Figure 15.1. Visual demonstration of the effects of shrinkage; each dashed line indicates participant-specific estimates, either based on (a) separate regression models for every individual (no pooling) or (b) a mixed model with varying intercepts and varying slopes; the bold line in (b) shows the population-level estimate of the mixed model; notice how, for the mixed model, intercepts and slopes are more similar to the population-level estimate

towards the intercept of the main population line. Similarly, the slopes are less extreme than in Figure 15.1a, closer to the slope of the population line. When compared to separate regressions, the individual-level estimates of a mixed model are drawn towards the average.

Gelman and Hill (2007: ch. 12) have a nice discussion of this. When a researcher ignores dependent groups of data points from the same individual (not including any random effects), they speak of 'complete pooling', where everything is just thrown together without teasing apart the dependencies. This approach tends to underestimate the variation present in the data. One may contrast this with a 'no-pooling' approach, with a separate model for each individual. This tends to result in more extreme estimates, often overestimating the variation across individuals. Mixed models can be thought of as a nice sweet spot between these two approaches: they neither ignore the individual level (as does 'complete pooling'), nor do they ignore the population level (as does 'no pooling').

This feature of mixed models makes these models ideal for studying individual differences. Drager and Hay (2012) discuss the utility of mixed models for studying individual differences in sociolinguistic behavior. Mirman, Dixon, and Magnuson (2008) discuss individual differences and mixed models in the context of eye-tracking research (see also Mirman et al., 2008; Mirman, 2014).

In my own research, I have found the random effects estimates to be often more theoretically interesting than the fixed effects estimates. For example, in a perception study on acoustic prominence (Baumann & Winter, 2018), we performed an exploratory analysis of individual differences based on the varying slopes estimates from a prior mixed model analysis. This analysis revealed different listener subgroups hidden in the random effects structure. In another study (Idemaru, Winter, Brown, & Oh, 2019), we compared the influence of pitch and loudness on politeness judgments. Investigating the random slopes revealed that all listeners interpreted the loudness manipulation the same way, whereas listeners varied greatly for the pitch manipulation. Thus, the random effects estimates from this study suggest that pitch is a less reliable acoustic cue to politeness than loudness.

15.10. Chapter Conclusions

This chapter has shown how to implement linear mixed effects models in R with lme4. You started by simulating data with by-participant and by-item dependencies. The conceptual pay-off of this exercise was that you could see how something as simple as duration measurements from vowels are influenced by a whole range of different sources of variation. The mixed model fitted on the simulated data allowed estimating the different variance components. You then created a bunch of messy models by either dropping important random effects or by asking the model to estimate things that aren't in the data, which led to estimation problems even though there were no warning messages. The take-home message of this section was that, no matter whether you obtain a warning message or not, it is important to investigate the random effects structure of your fitted model. You also learned how to derive p-values via likelihood ratio tests. Finally, after a discussion of convergence issues, you analyzed a dataset of the #uglyselfie phenomenon with mixed logistic regression and found that changing the optimizer facilitated solving a convergence problem. The chapter concluded with a brief discussion of shrinkage, and the utility of mixed effects models for studying individual differences.

15.11. Exercises

15.11.1. Exercise 1: Perform Likelihood Ratio Test

Perform a likelihood ratio test for the `Angle` predictor of the ugly selfies model. Do this first by specifying the null model yourself, using `anova()`. Then, do the same thing again with `afex`.

15.11.2. Exercise 2: Calculate Predicted Probabilities

For the #uglyselfie model, calculate the predicted probabilities for each camera angle condition. You may have to refer back to Chapter 12 on logistic regression.

16 Outlook and Strategies for Model Building

16.1. What You Have Learned So Far

This concluding chapter presents an overview of what you have learned in this book and discusses some recommendations for best practice in statistical modeling.

Let us take stock of your achievements. After learning about R (Chapter 1) and the tidyverse (Chapter 2), you have progressed to linear modeling. First, you modeled simple univariate distributions with means (Chapter 3), then means conditioned on another variable (Chapter 4). The following chapters delved more deeply into various aspects of the linear model framework, such as transforming your variables (Chapter 5), adding multiple predictors to your model (Chapter 6), as well as incorporating categorical predictors (Chapter 7) and interactions (Chapter 8). Throughout all of these chapters, you have almost exclusively looked at what your model predicts—this was a deliberate design choice for this course, as I wanted you to focus on interpreting your model before you engage with hypothesis testing, which followed in Chapters 9, 10, and 11. Up to Chapter 12, all response variables analyzed throughout the book were continuous. To model categorical data, Chapters 12 and 13 introduced the generalized linear model framework. You have learned about logistic regression to model binary responses (Chapter 12), and about Poisson regression to model count data (Chapter 13).

Finally, to model data with non-independent data points, I introduced you to mixed models (Chapters 14 and 15). Varying intercept models allow the intercepts to vary by a grouping factor (such as 'participant' or 'item'). Varying slope models additionally allow the slopes to vary by a grouping factor. The discussion highlighted that it's important not to default to any particular random effects structure, but to reason about which varying slopes are needed for specific fixed effects predictors in your study. Another take-home message was that you should never report the results of a non-converging model.

There's a clear trajectory through these chapters, with each chapter allowing you to fit more complex models. Moreover, as you progressed through the book, you learned how approaches in the earlier chapters were actually specific cases of larger frameworks. Specifically, linear regression models are a specific case of the generalized linear model (GLM) framework. And furthermore, GLMs are a specific case of the generalized linear mixed model framework—namely, they are models without random effects.

Throughout all of this, I emphasized that your models should be motivated by theory. This is easier said than done, and this section talks about the issues that revolve around choosing a good model.

16.2. Model Choice

A theme that runs through all of academic statistics is a tension between exploration and confirmation. Exploratory statistics is hypothesis-generating: finding novel relationships in data. Confirmatory statistics is hypothesis-testing: assessing the validity of pre-existing hypotheses with novel data. Both exploratory statistics and confirmatory statistics are necessary for scientific progress. Problems arise when confirmatory statistics are based on a prior exploration *of the same dataset.*

However, even when you actively try to use regression in a confirmatory fashion, the boundary between confirmation and exploration is not always that clear. Let's say you fit a model with a condition variable as one predictor, and gender as an additional control variable. Perhaps the gender variable turns out not to be significant, and you might be inclined to drop the variable in an effort to simplify the model. However, maybe after excluding gender from the model, the critical condition effect ceases to be significant. In such a situation, it is very difficult not to fall prey to making model choices that are guided by the significance of the condition variable. It is very easy to convince yourself after the fact that the model that included gender was actually the right one (for a discussion of these problems with linguistic examples, see Roettger, 2018).

The problem is that each of the explored models was essentially a new test of the same hypothesis. This amounts to a form of multiple testing and is thus bound to lead to an increase in Type I errors. Performing exploration and confirmation on the same dataset is dangerous and constitutes a form of data dredging (actively looking for significant effects; Simmons et al., 2011). It also leads to HARKing—hypothesizing after the results are known (Kerr, 1998). It's very easy to come up with an explanation for almost any phenomenon after the fact. Scientists are excellent at making sense of patterns, but when doing confirmatory statistics, the patterns have to be predicted in advance in order to count as a confirmation of a hypothesis.

The flexibility of the generalized linear mixed model framework is both a blessing and a curse. Perhaps we should aim to eradicate choices altogether? Wouldn't that be more 'objective'? In the following, I will discuss two choice-delimiting approaches. I will ultimately argue against both of them.

16.3. The Cookbook Approach

At several points in this book, I alluded to the difference between the 'statistical testing' and 'statistical modeling' mindsets. Statistical testing is what is often taught in introductory statistics classes, such as in undergraduate psychology courses. As part of this tradition, you may encounter recommendations such as the following (hypothetical quote):

If you want to test whether two groups differ in a continuous measure, use a *t*-test. If these observations are linked, use a paired (dependent samples) *t*-test, otherwise use an unpaired (independent samples) *t*-test. If you have more than two groups, use a one-way-ANOVA (analysis of variance). To add an additional condition variable, use a two-way ANOVA. To add a continuous covariate, use ANCOVA (analysis of co-variance).

This short paragraph is just a small subset of the possible tests one has to cover when following this approach. For example, there are also tests for categorical data (Chi-Square tests, binomial tests, etc.), as well as a plethora of non-parametric tests (Wilcoxon tests, Kruskal-Wallis tests, etc.). You quickly end up with a bewildering array of tests. Textbooks teaching this framework often assume what I call a 'cookbook approach', which involves teaching the reader a series of statistical tests, perhaps concluding with a flow chart that tells students how to choose the appropriate test for a given dataset.

When following this cookbook approach, the student ends up thinking more about what test to choose than about how to express their hypotheses using a statistical model. Each test is highly constraining, not giving the user much flexibility in expressing their theories. In the end, the cookbook approach results in a highly compartmentalized view of statistics.

In addition, I find that the cookbook approach actively discourages thinking about one's theory. When people use statistical tests such as t-tests and Chi-Square tests, they don't stop to think about what of their domain knowledge should be incorporated into their statistical models. Related to this, the whole perspective of making predictions is completely lost when using these methods. Ultimately, these tests have one clear goal, which is to generate a p-value.

In conclusion, the modeling framework has a host of advantages when compared to the testing framework.[1]

16.4. Stepwise Regression

By moving from statistical testing to statistical modeling, you gain access to a much more flexible toolkit. So, keeping with the theme of desperately trying to limit your choices, isn't there a way of automatically guiding your decisions about which model to fit? What about using an automated procedure that selects the best model for us?

Consider a situation with three predictors: A, B, and C. Ignoring interactions for the time being, you end up with the following possible models.

```
y ~ 1
y ~ 1 + A
y ~ 1 + B
y ~ 1 + C
y ~ 1 + A + B
y ~ 1 + A + C
y ~ 1 + B + C
y ~ 1 + A + B + C
```

---

1 That said, it may be useful for you to know how some of the tests that are commonly reported in the literature (such as t-tests, ANOVAs etc.) correspond to certain linear models discussed throughout this book. For this, see Appendix A.

A technique called 'stepwise regression' is one of many automatic procedures that can be used to select one model from the set of all possible models. Stepwise regression either performs 'forward selection' (starting with an empty model and iteratively adding predictors, keeping them in the model only if they are significant), 'backward selection' (starting with a full model and iteratively deleting those predictors that are not significant), or a combination of both. Stepwise regression is still quite common in linguistics, especially sociolinguistics.

Stepwise regression has the appearance of objectivity, but this approach has a host of problems that are well described in the statistical literature (Steyerberg, Eijkemans, & Habbema, 1999; Whittingham, Stephens, Bradbury, & Freckleton, 2006). One problem is that the final model resulting from a stepwise model selection procedure is often reported as the outcome of a confirmatory statistical analysis (with p-values and so on), as if it was predicted in advance. The procedure, however, performed a lot of exploration within the dataset to find the best model. As a result of this, Type I error rates for the final model are up through the roof, which has been verified by simulations (Mundry & Nunn, 2008; see also Steyerberg et al., 1999). This issue is related to the problem of multiple testing discussed in Chapter 10. Stepwise regression has the tendency to overfit, homing in on the idiosyncrasies of a particular dataset at hand. As a result, you have absolutely no guarantee that the final model generalizes well to novel data.[2] Even more problematic for our concerns, when performing stepwise regression, you essentially relegate all thinking to a machine, potentially ending up with a model that corresponds in no way to an established theory and that may move you far away from what you set out to investigate.

Thus, stepwise regression is fraught with a host of statistical and conceptual issues, which is why it is almost universally recommended against by statisticians. Simply put, linguists should not use stepwise regression. The only "problem" that stepwise regression solves is a lack of willingness to think deeply about one's models.

16.5. A Plea for Subjective and Theory-Driven Statistical Modeling

In this chapter, I have given you two examples of approaches that limit choices— either via highly specialized significance tests or via automatic model selection procedures such as stepwise regression. Instead, you should accept the fact that statistical modeling is subjective. McElreath (2016: 95) offers a nice quote for this:

> Making choices tends to make novices nervous. There's an illusion sometimes that default procedures are more objective than procedures that require user choice ... If that's true, then all "objective" means is that everyone does the same thing. It carries no guarantees of realism or accuracy.

---

2 There are also some more technical problems with stepwise regression. For example, forward and backward selection lead to different models. This is a conceptually thorny issue, as it means that different implementations result in different models. In addition, it's not entirely clear how to best perform stepwise regression in a mixed model context: should varying slopes be entered for each of the possible models? If varying slopes are added for all fixed effects, then this quickly leads to massively overspecified models that are bound to have convergence issues. If, on the other hand, varying slopes are not included, then estimates may be anti-conservative (Schielzeth & Forstmeier, 2009).

Statistical modeling requires deep thinking about one's theory and about the data. Ideally, each predictor you enter into a model has a strong reason for being included. Let's consider, for example, the predictor 'gender'. If you analyze sociolinguistic data, perhaps this variable should be included, as there is a lot of literature to suggest that women and men differ on a number of sociolinguistically relevant variables. For example, women have repeatedly been shown to be early adopters of sound changes. However, if you analyze psycholinguistic data, you may not need to include gender into the model, and you shouldn't do so if you have no specific hypothesis about gender differences for the phenomenon at hand. This is not to say that gender should never be a predictor in a model of psycholinguistic data. Quite the contrary: it should definitely be included in a model if there are specific theories that predict gender effects.

Consider an example: when we submitted the data that you analyzed in Chapter 6 (iconicity as a function of sensory experience, imageability, frequency, and systematicity), one of the reviewers asked us: "Why weren't any interactions between the variables fitted?" In our reviewer response letter, we provided a simple answer to this question: because we didn't predict any interactions. We had clear expectations for the individual effects of each predictor, and this is why they were included in the model in the first place. However, in the absence of strong theories about how these variables should interact, we deemed it was best not to try our luck. Thus, we decidedly stuck with the confirmatory fashion of our analysis, in spite of the reviewer's suggestion, which essentially just invited us to explore the data.

You can pre-empt problems at the modeling stage by specifying the model as much as possible in advance of your analysis, which is best done even before conducting the study. After all, if you are conducting a study, you are generally collecting data with a specific hypothesis in mind, which means that it should be possible to formulate your model formula before seeing the data. In fact, these days, more and more researchers pre-register their studies (such as via the Open Science Framework; see Nosek & Lakens, 2014), which includes publicly pre-specifying what analysis one should conduct.[3] This helps to be principled about making modeling decisions.

That said, it has to be recognized that experienced researchers find it easier than novices to make modeling decisions in advance of collecting the data. Novices will find it hard to foresee all the eventualities they may run into when performing an analysis. One's practice in statistical modeling is very much affected by experience—having seen lots of different datasets, having constructed lots of different residual plots, having encountered lots of convergence issues, and so on. As you advance in your career as a data analyst, you will find it progressively easier to make more principled decisions.

In doing statistical modeling, you should take a clear position on whether you are in 'confirmatory mode' or 'exploratory mode'. There is nothing wrong with exploratory statistics—on the contrary, exploration is extremely important. If it were not for exploration, many of the most important findings in science (including the language sciences) would have been missed. However, it is problematic when exploratory statistics are written up in the guise of confirmatory statistics. In such cases, it is best to clearly state that the analyses

3 Pre-registration is slowly but surely becoming institutionalized as more and more scientific journals adopt this practice. See http://cos.io/rr for updates and a growing list of journals (accessed October 23, 2018).

are exploratory, and that they therefore need to be confirmed with new data. In fact, part of the replication crisis that was discussed in Chapter 2 results from the fact that the research community seems to value confirmation more than exploration, which often requires reframing exploratory results as confirmatory in order to get them published. It's good to take a clear stance here and defend exploration when it is appropriate (Roettger, Winter, & Baayen, 2019).

Let me tell you about another example from my own research that I think nicely demonstrates the tension between exploration and confirmation. In Brown et al. (2014), we wanted to test whether Korean listeners can perceive politeness distinctions based on acoustic cues alone. We used mixed logistic regression in a confirmatory fashion to test this. However, we had no clear hypothesis about which specific acoustic cue would matter the most in determining politeness judgments. In fact, our own research (Winter & Grawunder, 2012) has shown that honorific distinctions correlate with a host of acoustic markers in production. Any one of them could have influenced politeness judgments in perception. In the absence of strong theoretically motivated predictions, we performed an exploratory analysis to investigate which acoustic cues were used by Korean listeners. In the paper, this analysis was clearly marked as exploratory, and we refrained from reporting any *p*-values. The patterns we found in this analysis need to be confirmed with new data. Moreover, our confirmatory mixed model analysis was fixed; we didn't change things as a result of the exploratory analysis. Thus, in this study, we followed McArdle's (2011: 335) mantra, "confirm first, explore second".

There's a thin line between exploration and confirmation. Both are incredibly important. However, when you're not clear on where you stand, and you start mixing exploration and confirmation on the same dataset, you enter the 'danger zone' of HARKing and data dredging.

16.6. Reproducible Research

The array of different options in statistical modeling potentially results in analyses where an outsider does not know which models have been explored prior to the model that is ultimately reported in a paper. This is the dark side of subjectivity. Reproducible research practices alleviate some of these concerns. Chapter 2 detailed many different ways you can make your analysis more open and accessible to others, which is a key ingredient to making it more reproducible. Ideally, all your data and code are publicly available, and ideally the code runs immediately on someone else's machine to reproduce exactly those numerical values that are reported in a paper.

There is a big push towards reproducible research in many different fields (Gentleman & Lang, 2007; Mesirov, 2010; Peng, 2011; Munafò et al., 2017), and linguistics is currently catching on (see Berez-Kroeker et al., 2018; Roettger, 2018; Roettger et al., 2019). Linguistics can only benefit from adopting these trends: not only does reproducibility allow other researchers to check the validity of published results, it also facilitates the exchange of methods and data. It's a way for the field to grow.

In one of his statistics workshops, John McArdle used the slogan "be honest, not pure". Of course, withdrawing from a desire to be 'pure' in one's statistical analyses should not be an invitation to 'sin' (e.g., fishing for effects, data dredging). However, this slogan reminds us that our models can never be perfect; they will never be able to satisfy all researchers. All we can do is to be honest about how our modeling decisions were made; we need to lay all our cards on the table. Because it is inevitable that different researchers will come to different conclusions even when given the same dataset (cf. Silberzahn et al., 2018), 'purity' is not an attainable goal. However, honesty is.

There simply is no 'best model' for a dataset that will satisfy the demands of all researchers. On a more personal note: looking back at some of my past models, I don't necessarily agree with all the decisions that I made. However, as everything is 'out there' for others to see (on GitHub and OSF), anybody who disagrees with my modeling decisions can form their own opinion. I hope that this book has encouraged you to be part of the growing movement towards reproducible research practices in the language sciences.

16.7. Closing Words

This book has covered a whole range of topics, but obviously there are many techniques not covered. However, regression (and its extensions) is one of the most useful tools to get you started. You will find that many of the new approaches you will hear about actually relate to regression in some form or another, although there are also many methods that are not based on this framework. Importantly, you should not see this book as the end point of your statistical journey. Instead, it should be a springboard for learning more. I hope to have prepared you for continuing your journey through the world of statistics.

References

Agresti, A. (2002). *Categorical data analysis*. Hoboken, NJ: John Wiley.

Auguie, B. (2017). gridExtra: Miscellaneous functions for "grid" graphics. R package version 2.3.

Austin, P.C., Mamdani, M.M., Juurlink, D.N., & Hux, J.E. (2006). Testing multiple statistical hypotheses resulted in spurious associations: A study of astrological signs and health. *Journal of Clinical Epidemiology*, *59*(9), 964–969.

Baayen, R.H. (2008). *Analyzing linguistic data*. Cambridge, UK: Cambridge University Press.

Baayen, R.H. (2013). languageR: Data sets and functions with "Analyzing linguistic data: A practical introduction to statistics". R package version 1.4.1.

Baayen, R.H., Davidson, D.J., & Bates, D.M. (2008). Mixed-effects modeling with crossed random effects for subjects and items. *Journal of Memory and Language*, *59*(4), 390–412.

Baayen, R.H., & Milin, P. (2010). Analyzing reaction times. *International Journal of Psychological Research*, *3*(2), 12–28.

Balota, D.A., Yap, M.J., Hutchison, K.A., Cortese, M.J., Kessler, B., Loftis, B., . . . Treiman, R. (2007). The English Lexicon Project. *Behavior Research Methods*, *39*, 445–459.

Barr, D.J., Levy, R., Scheepers, C., & Tily, J.J. (2013). Random-effects structure for confirmatory hypothesis testing: Keep it maximal. *Journal of Memory and Language*, *68*, 255–278.

Bartoń, K. (2017). MuMIn: Multi-model inference. R package version 1.40.0. Available online at https://cran.r-project.org/package=MuMIn

Bates, D., Maechler, M., Bolker, B., & Walker, S. (2015). Fitting linear mixed-effects models using lme4. *Journal of Statistical Software*, *67*(1), 1–48.

Baumann, S., & Winter, B. (2018). What makes a word prominent? Predicting untrained listeners' prominence judgments. *Journal of Phonetics*, *70*, 20–38.

Bednarek, M.A. (2008). Semantic preference and semantic prosody re-examined. *Corpus Linguistics and Linguistic Theory*, *4*, 119–139.

Bennett, C.M., Baird, A.A., Miller, M.B., & Wolford, G.L. (2011). Neural correlates of interspecies perspective taking in the post-mortem Atlantic salmon: an argument for proper multiple comparisons correction. *Journal of Serendipitous and Unexpected Results*, *1*, 1–5.

Bentz, C., & Winter, B. (2013). Languages with more second language learners tend to lose nominal case. *Language Dynamics & Change*, *3*(1), 1–27.

Berez-Kroeker, A.L., Gawne, L., Kung, S.S., Kelly, B.F., Heston, T., Holton, G., … Woodbury, A.C. (2018). Reproducible research in linguistics: A position statement on data citation and attribution in our field. *Linguistics*, *56*(1), 1–18.

Boot, I., & Pecher, D. (2010). Similarity is closeness: Metaphorical mapping in a conceptual task. *Quarterly Journal of Experimental Psychology, 63*, 942–954.

Breiman, L. (2001). Random forests. *Machine Learning, 45*, 5–32.

Bresnan, J., Cueni, A., Nikitina, T., & Baayen, R.H. (2007). Predicting the dative alternation. In G. Bouma, I. Kraemer, & J. Zwarts (Eds.), *Cognitive foundations of interpretation* (pp. 69–94). Amsterdam: Royal Netherlands Academy of Science.

Bresnan, J., & Hay, J. (2008). Gradient grammar: An effect of animacy on the syntax of *give* in New Zealand and American English. *Lingua, 118*(2), 245–259.

Brown, L., Winter, B., Idemaru, K., & Grawunder, S. (2014). Phonetics and politeness: Perceiving Korean honorific and non-honorific speech through phonetic cues. *Journal of Pragmatics, 66*, 45–60.

Brysbaert, M., & New, B. (2009). Moving beyond Kučera and Francis: A critical evaluation of current word frequency norms and the introduction of a new and improved word frequency measure for American English. *Behavior Research Methods, 41*, 977–990.

Brysbaert, M., New, B., & Keuleers, E. (2012). Adding part-of-speech information to the SUBTLEX-US word frequencies. *Behavior Research Methods, 44*(4), 991–997.

Brysbaert, M., & Stevens, M. (2018). Power analysis and effect size in mixed effects models: A tutorial. *Journal of Cognition, 1*(1), 9.

Bürkner, P.-C. (2017). brms: An R package for Bayesian multilevel models using Stan. *Journal of Statistical Software, 80*(1), 1–28.

Burns, P. (2011). *The R Inferno.* Available online: https://www.burns-stat.com/pages/Tutor/R_inferno.pdf

Buzsáki, G., & Mizuseki, K. (2014). The log-dynamic brain: How skewed distributions affect network operations. *Nature Reviews Neuroscience, 15*(4), 264–278.

Casasanto, D. (2008). Similarity and proximity: When does close in space mean close in mind? *Memory & Cognition, 36*, 1047–1056.

Christiansen, M.H., & Chater, N. (2016). *Creating language: Integrating evolution, acquisition, and processing.* Cambridge, MA: MIT Press.

Cleveland, W.S. (1984). Graphical methods for data presentation: Full scale breaks, dot charts, and multibased logging. *American Statistician, 38*(4), 270–280.

Cohen, J. (1988). *Statistical Power Analysis for the Behavioral Sciences* (2nd ed.). Hillsdale, NJ: Erlbaum Press.

Connell, L., & Lynott, D. (2012). Strength of perceptual experience predicts word processing performance better than concreteness or imageability. *Cognition, 125*(3), 452–465.

Cortese, M.J., & Fugett, A. (2004). Imageability ratings for 3,000 monosyllabic words. *Behavior Research Methods, Instruments, & Computers, 36*, 384–387.

Cumming, G. (2012). *Understanding the new statistics: Effect sizes, confidence intervals, and meta-analysis.* New York: Routledge.

Cumming, G. (2014). The new statistics: Why and how. *Psychological Science, 25*(1), 7–29.

Davies, M. (2008) The Corpus of Contemporary American English: 450 million words, 1990-present. Available online at http://corpus.byu.edu/coca/

de Bruin, A., Bak, T.H., & Della Sala, S. (2015). Examining the effects of active versus inactive bilingualism on executive control in a carefully matched non-immigrant sample. *Journal of Memory and Language, 85*, 15–26.

Dehaene, S. (2003). The neural basis of the Weber-Fechner law: A logarithmic mental number line. *Trends in Cognitive Sciences, 7*(4), 145–147.

Dienes, Z. (2008). *Understanding psychology as a science: An introduction to scientific and statistical inference.* New York: Palgrave Macmillan.

Dingemanse, M., Blasi, D.E., Lupyan, G., Christiansen, M.H., & Monaghan, P. (2015). Arbitrariness, iconicity, and systematicity in language. *Trends in Cognitive Sciences, 19*(10), 603–615.

Drager, K., & Hay, J. (2012). Exploiting random intercepts: Two case studies in sociophonetics. *Language Variation and Change, 24*(1), 59–78.

Eager, C., & Roy, J. (2017). Mixed effects models are sometimes terrible. Available online at https://arxiv.org/abs/1701.04858

Faraway, J. (2005). *Linear models with R.* Boca Raton, FL: Chapman & Hall/CRC Press.

Faraway, J.J. (2006). Extending the linear model with R: Generalized linear, mixed effects and nonparametric regression models. Boca Raton, FL: Chapman & Hall/CRC Press.

Fox, J., & Weisberg, S. (2011). *An R Companion to Applied Regression* (2nd ed.). Thousand Oaks CA: Sage.

Freeberg, T.M., & Lucas, J.R. (2009). Pseudoreplication is (still) a problem. *Journal of Comparative Psychology, 123*(4), 450–451.

García-Berthou, E., & Hurlbert, S.H. (1999). Pseudoreplication in hermit crab shell selection experiments: Comment to Wilber. *Bulletin of Marine Sciences, 65*(3), 893–895.

Gardner, M.J., & Altman, D.G. (1986). Confidence intervals rather than P values: Estimation rather than hypothesis testing. *British Medical Journal, 292*, 746–750.

Gasser, M. (2004). The origins of arbitrariness in language. In K. Forbus, D. Gentner, & T. Regier (Eds.), *Proceedings of the 26th annual conference of the Cognitive Science Society* (pp. 434–439). Mahwah, NJ: Erlbaum.

Gelman, A., & Carlin, J. (2014). Beyond power calculations: assessing type S (sign) and type M (magnitude) errors. *Perspectives on Psychological Science, 9*(6), 641–651.

Gelman, A., & Hill, J. (2007). *Data analysis using regression and multilevel/hierarchical models.* Cambridge, UK: Cambridge University Press.

Gelman, A., & Loken, E. (2014). Ethics and statistics: The AAA tranche of subprime science. *Chance, 27*(1), 51–56.

Gentleman, R., & Lang, D. (2007). Statistical analyses and reproducible research. *Journal of Computational and Graphical Statistics, 16*, 1–23.

Gigerenzer, G. (2004). Mindless statistics. *Journal of Socio-Economics, 33*(5), 587–606.

Gillespie, C., & Lovelace, R. (2017). *Efficient R programming.* Sebastopol, CA: O'Reilly.

Goodman, S.N. (1999). Toward evidence-based medical statistics. 1: The P value fallacy. *Annals of Internal Medicine, 130*(12), 995–1004.

Green, P., & MacLeod, C.J. (2016). SIMR: An R package for power analysis of generalized linear mixed models by simulation. *Methods in Ecology and Evolution, 7*(4), 493–498.

Gries, S.Th. (2015). The most under-used statistical method in corpus linguistics: Multilevel (and mixed-effects) models. *Corpora, 10*(1), 95–125.

Gudicha, D.W., Schmittmann, V.D., & Vermunt, J.K. (2017). Statistical power of likelihood ratio and Wald tests in latent class models with covariates. *Behavior Research Methods, 49*(5), 1824–1837.

Hassemer, J. (2016). Towards a theory of gesture form analysis. Imaginary forms as part of gesture conceptualisation, with empirical support from motion-capture data. PhD thesis, RWTH Aachen University.

Hassemer, J., & Winter, B. (2016). Producing and perceiving gestures conveying height or shape. *Gesture, 15*(3), 404–424.

Hassemer, J., & Winter, B. (2018). Decoding gestural iconicity. *Cognitive Science, 42*(8), 3034–3049.

Hauck Jr, W.W., & Donner, A. (1977). Wald's test as applied to hypotheses in logit analysis. *Journal of the American Statistical Association, 72*, 851–853.

Houtkoop, B.L., Chambers, C., Macleod, M., Bishop, D.V., Nichols, T.E., & Wagenmakers, E.J. (2018). Data sharing in psychology: A survey on barriers and preconditions. *Advances in Methods and Practices in Psychological Science, 1*(1), 70–85.

Hubbard, R., & Lindsay, R.M. (2008). Why P values are not a useful measure of evidence in statistical significance testing. *Theory & Psychology, 18*(1), 69–88.

Hunston, S. (2007). Semantic prosody revisited. *International Journal of Corpus Linguistics, 12*(2), 249–268.

Hurlbert, S.H. (1984). Pseudoreplication and the design of ecological field experiments. *Ecological Monographs, 54*(2), 187–211.

Idemaru, K., Winter, B., Brown, L., & Oh, G.E. (2019). Loudness trumps pitch in politeness judgments: Evidence from Korean deferential speech. *Language and Speech*. DOI: 10.1177/0023830918824344

Ioannidis, J.P. (2005). Why most published research findings are false. *PLoS Medicine, 2*(8), e124.

Jackman, S. (2015). pscl: Classes and methods for R developed in the Political Science Computational Laboratory, Stanford University. Department of Political Science, Stanford University. Stanford, California. R package version 1.4.9. Available online at http://pscl.stanford.edu/

Jaeger, T.F. (2008). Categorical data analysis: Away from ANOVAs (transformation or not) and towards logit mixed models. *Journal of Memory and Language, 59*(4), 434–446.

Jaeger, T.F., Graff, P., Croft, W., & Pontillo, D. (2011). Mixed effect models for genetic and areal dependencies in linguistic typology. *Linguistic Typology, 15*(2), 281–319.

Jescheniak, J.D., & Levelt, W.J. (1994). Word frequency effects in speech production: Retrieval of syntactic information and of phonological form. *Journal of Experimental Psychology: Learning, Memory, and Cognition, 20*, 824–843.

Juhasz, B.J., & Yap, M.J. (2013). Sensory experience ratings for over 5,000 mono-and disyllabic words. *Behavior Research Methods, 45*, 160–168.

Jurafsky, D. (2014). *The language of food.* New York: W.W. Norton.

Kello, C.T., Anderson, G.G., Holden, J.G., & Van Orden, G.C. (2008). The pervasiveness of 1/f scaling in speech reflects the metastable basis of cognition. *Cognitive Science, 32*(7), 1217–1231.

Kello, C.T., Brown, G.D., Ferrer-i-Cancho, R., Holden, J.G., Linkenkaer-Hansen, K., Rhodes, T., & Van Orden, G.C. (2010). Scaling laws in cognitive sciences. *Trends in Cognitive Sciences, 14*(5), 223–232.

Kerr, N.L. (1998). HARKing: Hypothesizing after the results are known. *Personality and Social Psychology Review, 2*(3), 196–217.

Kirby, J., & Sonderegger, M. (2018). Mixed-effects design analysis for experimental phonetics. *Journal of Phonetics, 70*, 70–85.

Kline, R.B. (2004). *Beyond significance testing: Reforming data analysis methods in behavioral research.* Washington, DC: American Psychological Association.

Krantz, D.H. (1999). The null hypothesis testing controversy in psychology. *Journal of the American Statistical Association, 94*(448), 1372–1381.

Krifka, M. (2010). A note on the asymmetry in the hedonic implicatures of olfactory and gustatory terms. In S. Fuchs, P. Hoole, C. Mooshammer & M. Zygis (Eds.), *Between the Regular and the Particular in Speech and Language* (pp. 235–245). Frankfurt am Main: Peter Lang.

Kroodsma, D. (1989). Suggested experimental designs for song playbacks. *Animal Behaviour, 37*, 600–609.

Kroodsma, D.E. (1990). Using appropriate experimental designs for intended hypotheses in "song" playbacks, with examples for testing effects of song repertoire sizes. *Animal Behaviour, 40*, 1138–1150.

Lazic, S.E. (2010). The problem of pseudoreplication in neuroscientific studies: Is it affecting your analysis? *BMC Neuroscience, 11*, 1–17.

Lenth, R. (2018). emmeans: Estimated marginal means, aka least-squares means. R package version 1.2.4.

Levinson, S.C., & Majid, A. (2014). Differential ineffability and the senses. *Mind & Language, 29*, 407–427.

Levy, R. (2018). Using R formulae to test for main effects in the presence of higher-order interactions. *arXiV*, 1405.2094v2. Available online at https://arxiv.org/pdf/1405.2094.pdf

Lievers, F.S., & Winter, B. (2018). Sensory language across lexical categories. *Lingua, 204*, 45–61.

Littlemore, J., Pérez Sobrino, P., Houghton, D., Shi, J., & Winter, B. (2018). What makes a good metaphor? A cross-cultural study of computer-generated metaphor appreciation. *Metaphor & Symbol, 33*, 101–122.

Lombardi, C.M., & Hurlbert, S.H. (1996). Sunfish cognition and pseudoreplication. *Animal Behaviour, 52*, 419–422.

Lynott, D., & Connell, L. (2009). Modality exclusivity norms for 423 object properties. *Behavior Research Methods, 41*, 558–564.

Lynott, D., & Connell, L. (2013). Modality exclusivity norms for 400 nouns: The relationship between perceptual experience and surface word form. *Behavior Research Methods, 45*(2), 516–526.

Machlis, L., Dodd, P.W.D., & Fentress, J.C. (1985). The pooling fallacy: Problems arising when individuals contribute more than one observation to the data set. *Zeitschrift für Tierpsychologie, 68J)*, 201–214.

Majid, A., & Burenhult, N. (2014). Odors are expressible in language, as long as you speak the right language. *Cognition, 130*, 266–270.

Matuschek, H., Kliegl, R., Vasishth, S., Baayen, H., & Bates, D. (2017). Balancing Type I error and power in linear mixed models. *Journal of Memory and Language, 94*, 305–315.

McArdle, J.J. (2011). Some ethical issues in factor analysis. In A.T. Panter & S.K. Sterba (Eds.), *Handbook of ethics in quantitative methodology* (pp. 313–339). New York: Routledge.

McElreath, R. (2016). *Statistical rethinking: A Bayesian course with examples in R and Stan*. Boca Raton, FL: CRC Press.

Mesirov, J. P. (2010). Computer science: Accessible reproducible research. *Science, 327*, 5964.

Milinski, M. (1997). How to avoid seven deadly sins in the study of behavior. *Advances in the Study of Behavior, 26*, 159–180.

Milton Bache, S., & Wickham, H. (2014). magrittr: A forward-pipe operator for R.R package version 1.5. Available online at https://CRAN.R-project.org/package=magrittr

Mirman, D. (2014). *Growth curve analysis and visualization using R*. Boca Raton, FL: CRC Press.

Mirman, D., Dixon, J.A., & Magnuson, J.S. (2008). Statistical and computational models of the visual world paradigm: Growth curves and individual differences. *Journal of Memory and Language, 59*(4), 475–494.

Monaghan, P., Shillcock, R.C., Christiansen, M.H., & Kirby, S. (2014). How arbitrary is English? *Philosophical Transactions of the Royal Society of London: Series B, Biological Sciences*, 369, 20130299.

Montgomery, D.C., & Peck, E.A. (1992). *Introduction to linear regression analysis*. New York: Wiley.

Morey, R.D., Hoekstra, R., Rouder, J.N., Lee, M.D., & Wagenmakers, E.J. (2016). The fallacy of placing confidence in confidence intervals. *Psychonomic Bulletin & Review*, *23*(1), 103–123.

Morrissey, M.B., & Ruxton, G.D. (2018). Multiple regression is not multiple regressions: The meaning of multiple regression and the non-problem of collinearity. *Philosophy, Theory, and Practice in Biology*, *10*(3).

Müller, K., & Wickham, H. (2018). tibble: Simple data frames. R package version 1.4.2. Available online at https://CRAN.R-project.org/package=tibble

Munafò, M.R., Nosek, B.A., Bishop, D.V., Button, K.S., Chambers, C.D., du Sert, N.P., … Ioannidis, J.P. (2017). A manifesto for reproducible science. *Nature Human Behaviour*, *1*, 0021.

Mundry, R., & Nunn, C.L. (2008). Stepwise model fitting and statistical inference: Turning noise into signal pollution. *American Naturalist*, *173*(1), 119–123.

Nakagawa, S. (2004). A farewell to Bonferroni: The problems of low statistical power and publication bias. *Behavioral Ecology*, *15*(6), 1044–1045.

Nakagawa, S., & Cuthill, I.C. (2007). Effect size, confidence interval and statistical significance: A practical guide for biologists. *Biological Review*, *82*, 591–605.

Nakagawa, S., & Schielzeth, H. (2013). A general and simple method for obtaining R2 from generalized linear mixed-effects models. *Methods in Ecology and Evolution*, 4, 133–142.

Nettle, D. (1999). *Linguistic Diversity*. Oxford: Oxford University Press.

Nicenboim, B., Roettger, T.B., & Vasishth, S. (2018). Using meta-analysis for evidence synthesis: The case of incomplete neutralization in German. *Journal of Phonetics*, *70*, 39–55.

Nicenboim, B., & Vasishth, S. (2016). Statistical methods for linguistic research: Foundational ideas—Part II. *Language and Linguistics Compass*, *10*(11), 591–613.

Nickerson, R.S. (2000). Null hypothesis significance testing: A review of an old and continuing controversy. *Psychological Methods*, *5*(2), 241–301.

Nieuwland, M.S., Politzer-Ahles, S., Heyselaar, E., Segaert, K., Darley, E., Kazanina, N., … Mézière, D. (2018). Large-scale replication study reveals a limit on probabilistic prediction in language comprehension. *eLife*, 7.

Nosek, B.A., & Lakens, D. (2014). Registered reports. *Social Psychology*, *45*, 137–141.

O'brien, R.M. (2007). A caution regarding rules of thumb for variance inflation factors. *Quality & Quantity*, *41*(5), 673–690.

O'Hara, R.B., & Kotze, D. J. (2010). Do not log-transform count data. *Methods in Ecology and Evolution*, *1*(2), 118–122.

Open Science Collaboration. (2015). Estimating the reproducibility of psychological science. *Science*, *349*(6251), aac4716.

Osborne, J. (2005). Notes on the use of data transformations. *Practical Assessment, Research and Evaluation*, *9*(1), 42–50.

Paap, K.R., & Greenberg, Z.I. (2013). There is no coherent evidence for a bilingual advantage in executive processing. *Cognitive Psychology*, *66*(2), 232–258.

Papesh, M.H. (2015). Just out of reach: On the reliability of the action-sentence compatibility effect. *Journal of Experimental Psychology: General*, *144*(6), e116–e141.

Peng, R.D. (2011). Reproducible research in computational science. *Science, 334*, 1226–1227.

Perezgonzalez, J.D. (2015). Fisher, Neyman-Pearson or NHST? A tutorial for teaching data testing. *Frontiers in Psychology, 6*, 223.

Perry, L.K., Perlman, M., & Lupyan, G. (2015). Iconicity in English and Spanish and its relation to lexical category and age of acquisition. *PloS One, 10*(9), e0137147.

Perry, L.K., Perlman, M., Winter, B., Massaro, D.W., & Lupyan, G. (2017). Iconicity in the speech of children and adults. *Developmental Science*, e12572.

Pinheiro, J.C., & Bates, D.M. (2000). *Mixed-effects models in S and SPLUS*. New York: Springer.

Piwowar, H.A., & Vision, T.J. (2013). Data reuse and the open data citation advantage. *PeerJ, 1*, e175.

Postman, K., & Conger, B. (1954). Verbal habits and the visual recognition of words. *Science, 119*, 671–673.

Quinn, G.P., & Keough, M.J. (2002). *Experimental design and data analysis for biologists*. Cambridge, UK: Cambridge University Press.

Reinboud, W. (2004). Linear models can't keep up with sport gender gap. *Nature, 432*(7014), 147.

Reinhart, C.M., & Rogoff, K.S. (2010). Growth in a time of debt. *American Economic Review, 100*(2), 573–578.

Rice, K. (2004). Sprint research runs into a credibility gap. *Nature, 432*(7014), 147.

Roberts, S., & Winters, J. (2013). Linguistic diversity and traffic accidents: Lessons from statistical studies of cultural traits. *PloS One, 8*(8), e70902.

Robinson, D. (2017). broom: Convert statistical analysis objects into tidy data frames. R package version 0.4.3. Available online at https://CRAN.R-project.org/package=broom

Roettger, T.B. (2018). Researcher degrees of freedom in phonetic research. *Journal of the Association for Laboratory Phonology, 10*(1).

Roettger, T.B., Winter, B., & Baayen, R.H. (2019). Emergent data analysis in phonetic sciences: Towards pluralism and reproducibility. *Journal of Phonetics, 73*, 1–7.

Roettger, T.B., Winter, B., Grawunder, S., Kirby, J., & Grice, M. (2014). Assessing incomplete neutralization of final devoicing in German. *Journal of Phonetics, 43*, 11–25.

Rothman, K.J. (1990). No adjustments are needed for multiple comparisons. *Epidemiology, 1*(1), 43–46.

Rouby, C., & Bensafi, M. (2002). Is there a hedonic dimension to odors? In C. Rouby, B. Schaal, D. Dubois, R. Gervais, & A. Holley (Eds.), *Olfaction, taste, and cognition* (pp. 140–159). Cambridge, UK: Cambridge University Press.

Schiel, F., Heinrich, C., & Barfüsser, S. (2012). Alcohol language corpus: The first public corpus of alcoholized German speech. *Language Resources and Evaluation, 46*(3), 503–521.

Schielzeth, H. (2010). Simple means to improve the interpretability of regression coefficients. *Methods in Ecology and Evolution, 1*(2), 103–113.

Schielzeth, H., & Forstmeier, W. (2009). Conclusions beyond support: Overconfident estimates in mixed models. *Behavioral Ecology, 20*, 416–420.

Seedorff, M., Oleson, J., & McMurray, B. (2019). Maybe maximal: Good enough mixed models optimize power while controlling Type I error. PsyArXiv pre-print, DOI: 10.31234/osf.io/xmhfr Available online at https://psyarxiv.com/xmhfr/

Shaoul, C., & Westbury, C. (2010). Exploring lexical co-occurrence space using HiDEx. *Behavior Research Methods, 42*(2), 393–413.

Sidhu, D.M., & Pexman, P.M. (2018). Lonely sensational icons: Semantic neighbourhood density, sensory experience and iconicity. *Language, Cognition and Neuroscience, 33*(1), 25–31.

Silberzahn, R., Uhlmann, E. L., Martin, D. P., Anselmi, P., Aust, F., Awtrey, E., ... Carlsson, R. (2018). Many analysts, one data set: Making transparent how variations in analytic choices affect results. *Advances in Methods and Practices in Psychological Science, 1*(3), 337–356.

Simmons, J.P., Nelson, L.D., & Simonsohn, U. (2011). False-positive psychology: Undisclosed flexibility in data collection and analysis allows presenting anything as significant. *Psychological Science, 22*(11), 1359–1366.

Singmann, H., Bolker, B., Westfall, J., & Aust, F. (2016). afex: Analysis of factorial experiments. R package version 0.16–1. Available online at https://CRAN.R-project.org/package=afex

Smith, N.J., & Levy, R. (2013). The effect of word predictability on reading time is logarithmic. *Cognition, 128*(3), 302–319.

Snefjella, B., & Kuperman, V. (2016). It's all in the delivery: Effects of context valence, arousal, and concreteness on visual word processing. *Cognition, 156*, 135–146.

Solomon, R.L., & Postman, L. (1952). Frequency of usage as a determinant of recognition thresholds for words. *Journal of Experimental Psychology, 43*, 195–201.

Sóskuthy, M. (2017). Generalised additive mixed models for dynamic analysis in linguistics: A practical introduction. arXiv preprint arXiv:1703.05339. Available online at http://eprints.whiterose.ac.uk/113858/2/1703_05339v1.pdf

Stack, C.M.H., James, A.N., & Watson, D.G. (2018). A failure to replicate rapid syntactic adaptation in comprehension. *Memory & Cognition, 46*(6), 864–877.

Sterne, J.A., & Smith, G.D. (2001). Sifting the evidence—What's wrong with significance tests? *Physical Therapy, 81*(8), 1464–1469.

Stevens, S.S. (1957). On the psychophysical law. *Psychological Review, 64*(3), 153–181.

Steyerberg, E.W., Eijkemans, M.J., & Habbema, J.D.F. (1999). Stepwise selection in small data sets: A simulation study of bias in logistic regression analysis. *Journal of Clinical Epidemiology, 52*(10), 935–942.

Strobl, C., Malley, J., & Tutz, G. (2009). An introduction to recursive partitioning: Rationale, application, and characteristics of classification and regression trees, bagging, and random forests. *Psychological Methods, 14*(4), 323–348.

Tagliamonte, S.A., & Baayen, H. (2012). Models, forests, and trees of York English: *Was/were* variation as a case study for statistical practice. *Language Variation and Change, 24*(2), 135–178.

Tatem, A.J., Guerra, C.A., Atkinson, P.M., & Hay, S.I. (2004). Athletics: Momentous sprint at the 2156 Olympics? *Nature, 431*(7008), 525.

Thompson, B. (2004). The "significance" crisis in psychology and education. *Journal of Socio-Economics, 33*(5), 607–613.

Tomaschek, F., Hendrix, P., & Baayen, R.H. (2018). Strategies for addressing collinearity in multivariate linguistic data. *Journal of Phonetics, 71*, 249–267.

Torchiano, M. (2016). *effsize: Efficient effect size computation*. R package version 0.6.4.

Vasishth, S., & Gelman, A. (2017). The statistical significance filter leads to overconfident expectations of replicability. arXiv preprint arXiv:1702.00556. Available online at www.stat.columbia.edu/~gelman/research/unpublished/VasishthGelmanCogSci2017.pdf

Vasishth, S., & Nicenboim, B. (2016). Statistical methods for linguistic research: Foundational ideas—Part I. *Language and Linguistics Compass, 10*(8), 349–369.

Vasishth, S., Nicenboim, B., Beckman, M.E., Li, F., Kong, E.-J. (2018). Bayesian data analysis in the phonetic sciences: A tutorial introduction. *Journal of Phonetics, 71*, 147–161.

Venables, W.N., & Ripley, B.D. (2002). *Modern applied statistics with S.* (4th ed.). New York: Springer.

Vinson, D.W., & Dale, R. (2014). Valence weakly constrains the information density of messages. In P. Bello, M. Guarini, M. McShane, & B. Scassellati (Eds.) *Proceedings of* the 36th annual meeting of the Cognitive Science Society (pp. 1682–1687). Austin, TX: Cognitive Science Society.

Warriner, A.B., Kuperman, V., & Brysbaert, M. (2013). Norms of valence, arousal, and dominance for 13,915 English lemmas. *Behavior Research Methods, 45*, 1191–1207.

Whittingham, M.J., Stephens, P.A., Bradbury, R.B., & Freckleton, R.P. (2006). Why do we still use stepwise modelling in ecology and behaviour? *Journal of Animal Ecology, 75*(5), 1182–1189.

Wickham, H. (2016). *ggplot2: Elegant graphics for data analysis*. New York: Springer-Verlag, 2016.

Wickham, H. (2017). tidyverse: Easily install and load the "tidyverse". R package version 1.2.1. Available online at https://CRAN.R-project.org/package=tidyverse

Wickham, H., François, R., Henry, L., & Müller, K. (2018). dplyr: A grammar of data manipulation. R package version 0.7.5. Available online at https://CRAN.R-project.org/package=dplyr

Wickham, H., Hester, J., & François, R. (2017). readr: Read rectangular text data. R package version 1.1.1. Available online at https://CRAN.R-project.org/package=readr

Wickham, H., & Grolemund, G (2017). *R for data science*. Sebastopol, CA: O'Reilly.

Wieling, M. (2018). Analyzing dynamic phonetic data using generalized additive mixed modeling: A tutorial focusing on articulatory differences between L1 and L2 speakers of English. *Journal of Phonetics, 70*, 86–116.

Williamson, J.M., Lin, H., Lyles, R.H., & Hightower, A.W. (2007). Power calculations for ZIP and ZINB models. *Journal of Data Science, 5*, 519–534.

Winter, B. (2011). Pseudoreplication in phonetic research. *Proceedings of the International Congress of Phonetic Science* (pp. 2137–2140). Hong Kong, August 17–21, 2011.

Winter, B. (2016). Taste and smell words form an affectively loaded part of the English lexicon. *Language, Cognition and Neuroscience, 31*(8), 975–988.

Winter, B., & Bergen, B. (2012). Language comprehenders represent object distance both visually and auditorily. *Language and Cognition, 4*(1), 1–16.

Winter, B., & Grawunder, S. (2012). The phonetic profile of Korean formality. *Journal of Phonetics, 40*(6), 808–815.

Winter, B., & Matlock, T. (2013). Making judgments based on similarity and proximity. *Metaphor & Symbol, 28*, 219–232.

Winter, B., Perlman, M., & Majid, A. (2018). Vision dominates in perceptual language: English sensory vocabulary is optimized for usage. *Cognition, 179*, 213–220.

Winter, B., Perlman, M., Perry, L.K., & Lupyan, G. (2017). Which words are most iconic? Iconicity in English sensory words. *Interaction Studies, 18*(3), 433–454.

Winter, B., & Wieling, M. (2016). How to analyze linguistic change using mixed models, Growth Curve Analysis and Generalized Additive Modeling. *Journal of Language Evolution, 1*, 7–18.

Xie, Y. (2015). *Dynamic documents with R and knitr*. Boca Raton, FL: Chapman and Hall/ CRC Press.

Xie, Y. (2018). knitr: A general-purpose package for dynamic report generation in R. R package version 1.20. Available online at https://cran.r-project.org/ packages=knitr

Zipf, G.K. (1949). *Human behavior and the principle of least effort: An introduction to human ecology*. Reading, MA: Addison Wesley.

Zuur, A.F., Ieno, E.N., & Elphick, C.S. (2010). A protocol for data exploration to avoid common statistical problems. *Methods in Ecology and Evolution, 1*(1), 3–14.

Zuur, A.F., Ieno, E.N., Walker, N.J., Saveliev, A.A., & Smith, G.M. (2009). *Mixed effects models and extensions in ecology with R*. New York: Springer.

Appendix A
Correspondences Between Significance Tests and Linear Models

This appendix serves two purposes. First, if you already know basic significance tests such as the *t*-test it will help you to understand how these tests map onto the corresponding linear models. Second, if you do not know these tests, then this appendix serves as a brief introduction.

I recommend reading this appendix only if you have already completed Chapters 1 to 7, as well as Chapter 9 (significance testing). In addition, I recommend having a look at Chapter 16 first.

Load the `tidyverse` and `broom` packages before beginning the chapter:

```
library(tidyverse)

library(broom)
```

A1. *t*-Tests

First, let us focus on what is perhaps the most commonly discussed significance test, the *t*-test. For the *t*-test, the response has to be continuous. For example, you may be interested in whether there is a difference in voice pitch between women and men (see Chapter 9), or whether there is a difference in the emotional valence between taste and smell words (see Chapter 7).

Let us start by creating some data in R. For now, let's work with the example of voice pitch, which is the perceptual correlate of fundamental frequency (how quickly your vocal folds vibrate). Voice pitch is measured on the continuous scale of Hertz. The following code creates two vectors, M and F. Each vector includes a set of 50 random numbers that are drawn from the normal distribution with the `rnorm()` function. The means of the respective groups are specified to be 200 Hz for women and 100 Hz for men. The standard deviation for both groups is specified to be 10 Hz.

```
F <- rnorm(50, mean = 200, sd = 10) # female values

M <- rnorm(50, mean = 100, sd = 10) # male values
```

Next, let's combine these two vectors into one vector, using the concatenate function `c()`. In the resulting vector, all male values are listed after all female values.

```
resp <- c(F, M) # concatenate both
```

Let's create gender identifiers. For this, concatenate the two character labels 'F' and 'M' together with the `c()` function. Then, take the resulting vector and instruct the repeat function `rep()` to repeat each one of the concatenated labels 50 times.

```
gender <- rep(c('F', 'M'), each = 50) # create gender ids
```

Next, put both into a tibble called `df`.

```
df <- tibble(gender, resp)
```

Let's quickly check how the tibble looks like (remember that your numbers will be different due to random sampling).

```
df
```

```
# A tibble: 100 x 2
   gender   resp
   <chr>    <dbl>
 1 F        214.
 2 F        194.
 3 F        204.
 4 F        206.
 5 F        204.
 6 F        199.
 7 F        215.
 8 F        199.
 9 F        220.
10 F        199.
# ... with 90 more rows
```

Now that you have a tibble with the data in place, you can perform a *t*-test to establish whether the two groups are significantly different from each other. This corresponds to the following logic (see Chapter 9): assuming that female and male voice pitches are equal (= null hypothesis of 0 difference), how probable is the actually observed difference or any difference more extreme than that? This is how you can perform the test in R (explanations follow).

```
t.test(resp ~ gender, data = df,
       paired = FALSE, var.equal = TRUE)
```

```
    Two Sample t-test

data: resp by gender
t = 47.222, df = 98, p-value < 2.2e-16
```

```
alternative hypothesis: true difference in means is not
equal to 0
95 percent confidence interval:
94.49116 102.78137
sample estimates:
mean in group F mean in group M
       199.6433        101.0070
```

You set the argument `paired` = `FALSE` because you have two independent groups of data points, one set of female pitch values and one set of male pitch values, what is called an 'unpaired *t*-test' or 'independent samples *t*-test'. The argument `var.equal` = `TRUE` specifies that you assume the variances in both groups to be equal (think of the homoscedasticity assumption discussed in Chapter 4). You can safely do this in this case because you specified the standard deviations to be equal when the data was generated.

The *p*-value is very small. So, operating at an alpha level of α = 0.05, this dataset can be seen as sufficiently incompatible with the null hypothesis. In other words, this result is 'significant'. The linear model corresponding to an unpaired *t*-test simply looks like this:

```
xmdl <- lm(resp ~ gender, data = df)

tidy(xmdl)
```

```
          term estimate std.error statistic      p.value
1 (Intercept) 199.64328 1.476987 135.16927 3.439977e-113
2     genderM -98.63627 2.088775 -47.22206  3.184096e-69
```

As discussed in Chapter 7, categorical factors, such as (in this case) `gender`, are treatment-coded by default. The coefficient for `gender` then represents the difference between two groups, which is tested against 0. Barring some rounding differences due to differences in display defaults, the statistical results of `lm()` and `t.test()` are equivalent. In particular, notice that the test statistic is indicated to be t = 47.22 in both cases. The fact that it is negative in the case of the linear model is irrelevant; this merely depends on which group is subtracted first.

Next in the line of basic significance tests is the 'one-sample *t*-test'. This is a test where just one set of numbers (one sample) is tested against some pre-established number. For example, you have already been exposed to the iconicity ratings collected by Perry et al. (2015, 2017) and Winter et al. (2017). In these studies, we used a centered rating scale for iconicity that ranged from –5 ('the word sounds like the opposite of what it means') to +5 ('the word sounds like what it means'). In these studies, we reported a result where the mean of the overall distribution of iconicity ratings is tested against 0. The one-sample *t*-test can be used to achieve this, and it can also be used to test a set of numbers against any other value from the literature.

Just to gain some experience with this test, you can test whether the voice pitches just generated are reliably different from 0. Not a particularly interesting result, since fundamental frequencies have to be positive anyway.

```
# One-sample t-test:

t.test(resp, data = df, mu = 0)
```

```
        One Sample t-test

data: resp
t = 29.683, df = 99, p-value < 2.2e-16
alternative hypothesis: true mean is not equal to 0
95 percent confidence interval:
140.2763 160.3740
sample estimates:
mean of x
150.3251
```

The corresponding linear model is an intercept-only model:

```
# One-sample t-test with lm():

xmdl <- lm(resp ~ 1, data = df)

tidy(xmdl)
```

```
          term estimate std.error statistic     p.value
1 (Intercept) 150.3251  5.064405  29.68269 4.39096e-51
```

As discussed in Chapter 4, in the absence of any conditioning variables, a linear model simply predicts the mean of a dataset. The intercept is then tested against 0, which is often not an interesting comparison to make.[1] Comparing the output of lm() and t.test() shows that the reported statistics are the same, barring some rounding differences.

The last *t*-test to discuss is the 'paired *t*-test', otherwise known as 'dependent samples *t*-test'. This test is used when observations are linked, such as when each participant is exposed to two conditions. For example, if you wanted to know whether a group of participants improved after receiving some form of training, each participant would have a pre-test and a post-test score associated with them. It is in situations like this that a paired *t*-test is appropriate.

To create a dataset amenable to a paired *t*-test analysis, let's change the example. In the following pipeline, the gender column is first renamed to cond for 'condition'. Then, ifelse() is used to change the 'M' labels to 'post' (post-test), and the 'F' labels to 'pre' (pre-test). In the final step, the response is multiplied by 4 to make the example look like response durations. Let's say you're interested in testing whether participants speed up or slow down on some task after having received training.

---

1 If you wanted to test against another value (e.g., a particular value taken from the literature), the t.test() function allows specifying the mu argument.

```
df <- rename(df, cond = gender) %>%
  mutate(cond = ifelse(cond == 'M', 'post', 'pre'),
         resp = 4 * resp)
df
```

```
# A tibble: 100 x 2
   cond    resp
   <chr>   <dbl>
 1 pre     855.
 2 pre     777.
 3 pre     815.
 4 pre     825.
 5 pre     816.
 6 pre     796.
 7 pre     860.
 8 pre     796.
 9 pre     881.
10 pre     797.
# ... with 90 more rows
```

Now everything is in place for running a paired t-test:

```
t.test(resp ~ cond, df, paired = TRUE)
```

```
        Paired t-test

data: resp by cond
t = -42.52, df = 49, p-value < 2.2e-16
alternative hypothesis: true difference in means is not
equal to 0
95 percent confidence interval:
-413.1921 -375.8980
sample estimates:
mean of the differences
              -394.5451
```

Thus, there is a statistically significant difference between the pre-test and the post-test condition. How do you conduct a paired t-test in the linear model framework? For this it helps to understand that a paired t-test is actually just a one-sample t-test that tests the differences between two groups against 0. So, you can compute difference scores (post minus pre) and fit an intercept-only model. The following code achieves this:

```
posts <- filter(df, cond == 'post')$resp

pres <- filter(df, cond == 'pre')$resp

diffs <- posts - pres
```

```
xmdl <- lm(diffs ~ 1)
```

```
tidy(xmdl)
```

```
      term       estimate  std.error  statistic       p.value
1 (Intercept)   -394.5451  9.279103  -42.51974  2.423957e-40
```

As discussed in Chapter 4, an intercept-only model predicts the mean, so this model predicts the mean differences. The significance test of the intercept is then a test of the mean difference against 0, which is exactly the same calculation that a paired *t*-test performs.

Finally, let me emphasize that the *t*-test is fairly limited in its domain of application because it *assumes independence*. For an unpaired *t*-test, this means that each data point has to come from a different participant. For a paired *t*-test, this means that each participant can maximally contribute one pair of data points. This is why the linear model framework is advantageous, as it allows extending to mixed models to deal with cases where *t*-tests are too constrained.

A2. Tests for Categorical Data

The *t*-test is for continuous data (one group of numbers against an established mean, or two groups). For categorical data, there are various tests available, of which we will cover only the binomial test and the chi-square test ('χ^2-test'). Again, let's start by generating some random data. Remember that you will get slightly different results from what is reported in this book.

To generate 50 binomially distributed random numbers, use rbinom(). Let's set the probability argument prob to 0.8 and the size argument to 1. Thus, altogether, the command below generates 50 random numbers for which there is an 80% chance of observing an event (1), compared to a 20% chance of not observing it (0).

```
x <- rbinom(50, size = 1, prob = 0.8)
```

Let's check what's contained in the object x (remember: your numbers will be different).

```
x
```

```
 [1] 0 0 1 0 1 1 1 1 1 1 1 1 1 0 1 1 0 0 1 1 1 0 1 0
[24] 0 1 1 1 0 1 0 1 0 1 1 1 0 1 1 0 1 1 1 1 0 1 0
[47] 0 1 0 1
```

```
xtab <- table(x)
```

```
xtab
```

```
 0  1
12 38
```

One significance test you might want to perform here is for testing whether the probability of observing an event is significantly different from $p = 0.5$. This is what's called a 'binomial test'.

```
binom.test(xtab, p = 0.5)
```

```
      Exact binomial test
data: xtab
number of successes = 18, number of trials =
50, p-value = 0.06491
alternative hypothesis: true probability of success is not
equal to 0.5
95 percent confidence interval:
0.2291571 0.5080686
sample estimates:
probability of success
            0.36
```

The p-value is above 0.05 in this particular case, which means that, given this data, there is not enough evidence to refute the null hypothesis of equal proportions. This, of course, only applies to this specific dataset. For a different random dataset, you may obtain a significant result.

Binomial tests are extremely limited, but nonetheless useful in many circumstances. For example, in the case of the height versus shape gestures (see Chapter 12; Hassemer & Winter, 2016), you could use a binomial test to show that there are significantly more shape responses in the dataset than height responses. In Chapter 12, it was mentioned that we observed a total of 184 shape responses, and 125 height responses, in our gesture perception study. Plugging these two numbers into a binomial test shows that these counts are incompatible with the null hypothesis that the two response options have equal probability.

```
binom.test(c(125, 184))
```

```
      Exact binomial test

data: c(125, 184)
number of successes = 125, number of trials
= 309, p-value = 0.0009383
alternative hypothesis: true probability of success is not
equal to 0.5
95 percent confidence interval:
0.3493380 0.4615722
sample estimates:
probability of success
            0.4045307
```

Another test that often comes up in the context of categorical data is the chi-square test. Let's start with the chi-square test of equal proportions. As was mentioned in previous chapters, Lynott and Connell (2009) performed a sensory modality rating study for 423 English adjectives. There are 205 sight words, 70 touch words, 68 sound words, 54 taste words, and 26 smell words in their dataset. Do these counts differ from chance expectations? For this, you can conduct a chi-square test of equal proportions.

```
mods <- c(205, 70, 68, 54, 26)
```

```
chisq.test(mods)
```

```
        Chi-squared test for given probabilities
data: mods
X-squared = 228.78, df = 4, p-value <
2.2e-16
```

Thus, assuming that there is an equal number of words for all of the senses, the actually observed counts are fairly unexpected. You can look at the counts expected under the null hypothesis of equal proportions as follows:

```
chisq.test(mods)$expected
```

```
[1] 84.6 84.6 84.6 84.6 84.6
```

... which is the same as:

```
sum(mods) / length(mods)
```

```
[1] 84.6
```

In other words, you expect a count of 84.6 for each cell when assuming that the null hypothesis of equal proportions is true. The chi-square test tells you that the observed counts significantly deviate from these expected counts.

A more complex chi-square test can be exemplified with an unpublished study that I conducted with my former PhD supervisor Teenie Matlock. In this study, we wanted to know whether verbs of perception implicitly encode distance. For example, the sentence *You are looking at the door* seems to imply a distance farther away from the door than *You are inspecting the door*. We asked participants to draw the door described in each one of these sentences. Then, research assistants coded for whether the door included visual detail or not (such as a doorknob, a frame, a window, etc.). This was treated as a binary variable. We predicted that participants would draw more detail if the perception verb implies a closer distance to the door, since details can be spotted only when one is close to an object. So we want to know whether the binary response variable 'detail' ('yes' versus 'no') is affected by the binary predictor variable 'distance' ('near' versus 'far'). Let's load in the data:

```
xdist <- read_csv('winter_matlock_unpublished_distance.
csv')
```

```
xdist
```

```
# A tibble: 398 x 3
   subject Details Distance
```

```
      <chr>    <chr>    <chr>
 1 S1       Yes      near
 2 S2       Yes      near
 3 S3       Yes      far
 4 S4       Yes      far
 5 S5       Yes      near
 6 S6       Yes      far
 7 S7       Yes      near
 8 S8       No       far
 9 S9       No       far
10 S10      Yes      far
# ... with 388 more rows
```

Let's tabulate the contents of the `Distance` column against the contents of the `Details` column. In the following command, the `with()` function is used so that it isn't necessary to type the name of the tibble again when indexing column. The `with()` function takes two arguments: first, a tibble; second, a function. The contents of the tibble are then made available to the function.

```
xtab <- with(xdist, table(Distance, Details))

xtab
```

```
          Details
Distance  No Yes
    far   89 117
    near  59 133
```

To compute a chi-square test, wrap the `chisq.test()` function around the table.

```
chisq.test(xtab)
```

```
        Pearson's Chi-squared test with Yates' continuity
correction
data: xtab
X-squared = 6.0975, df = 1, p-value = 0.0135
```

So, a chi-square test performed on this 2 x 2 contingency table is 'significant' ($p <$ 0.05). What does 'significance' mean in the context of a two-dimensional table? To understand what's going on, you can look at the expected counts:

```
chisq.test(xtab)$expected
```

```
          Details
Distance       No        Yes
    far   76.60302  129.397
    near  71.39698  120.603
```

These are the counts that are expected under the null hypothesis *that the columns are independent from the rows*. In this particular case, the chi-square test assesses whether differences in `Distance` are associated with differences in `Details`. The particular chi-square test executed here bears the name 'chi-square test of independence'. Don't confuse this with the independence assumption discussed in Chapter 14. In fact, the chi-square test *assumes* independence—that is, each participant can at most contribute one data point. In other words, there cannot be multiple data points from the same individual within a given cell.

The expected counts are derived by taking the row totals and multiplying them by the column totals.[2] This can be paraphrased as follows: our expectation for a given cell is based on what row/column combination you are in. You expect more data in a cell if it is also in a row that has a lot of data, and the same applies to columns. Cells that deviate strongly from the expected counts contribute to a significant chi-square value.

To compute all of this with the corresponding linear model, you need logistic regression. For this, you first need to make the response into a factor (see Chapter 12).

```
xdist <- mutate(xdist, Details = factor(Details))


dist_mdl <- glm(Details ~ Distance,
    data = xdist, family = 'binomial')


tidy(dist_mdl)
          term   estimate std.error statistic    p.value
1  (Intercept) 0.2735376 0.1406519 1.944784 0.05180102
2 Distancenear 0.5392741 0.2103590 2.563590 0.01035959
```

The coefficient table shows a significant effect for the distance factor, with a positive log odd coefficient for the 'near' condition. These log odds indicate that when a sentence was 'near', there was a higher probability of observing a drawing with detail ('yes'). Table A1 lists the correspondences discussed up to this point.

A3. Other Tests

The set of tests shown in Table A1 is just a small subset of the most basic tests that can be re-expressed in linear model format. Another very frequent type of significance testing procedure is ANOVA, analysis of variance. However, the linear model can perform the same job as ANOVA.[3] If you are familiar with ANOVA, Table A2 maps some common ANOVAs onto the corresponding linear models.

---

2 Advanced R tip: here's how to calculate this by hand using the cross-multiplication function `outer()`:
```
outer(rowSums(xtab), colSums(xtab)) / sum(xtab)
          No      Yes
far   76.60302 129.397
near  71.39698 120.603
```
3 Depending on what the default settings of particular functions are, the exact numerical output may differ (e.g., depending on what types of sums of squares are computed).

Table A1. Correspondences between some significance tests and linear models

Significance test	Linear model	Description
t.test(y ~ pred, paired = FALSE)	lm(y ~ pred)	An unpaired *t*-test corresponds to a linear model with a binary categorical predictor
t.test(y, mu = 0)	lm(y ~ 1)	One-sample *t*-test corresponds to an intercept-only model
t.test(y ~ pred, paired = TRUE)	lm(diffs ~ 1)	A paired *t*-test corresponds to an intercept-only model fitted on differences
chisq.test(xtab)	glm(y ~ x, family = binomial)	A chi-square test can be emulated with a logistic regression model

Table A2. Correspondences between ANOVAs and linear models

ANOVA	Linear model	Description
aov(y ~ c3)	lm(y ~ c3)	One-way ANOVA with three-level factor
aov(y ~ c2 * c2)	lm(y ~ c2 * c2)	2 x 2 ANOVA (two-way ANOVA)
aov(y ~ c2 * c3)	lm(y ~ c2 * c3)	2 x 3 ANOVA (and so on)
aov(y ~ c2 * covariate)	lm(y ~ c2 * covariate)	ANCOVA (analysis of covariance) with covariate (continuous predictor) and many other types of similar models

Instead of repeated measures ANOVA, you can use mixed models. The linear model framework allows much more complex random effects structures, thus giving the user more flexibility in expressing their theories.

Appendix B
Reading Recommendations

How to continue learning after this book? This appendix contains a (very personal) list of reading recommendations. I have always found it helpful to mix very easy reads ('bedtime reads') with intermediate and more advanced reads. If you get stuck in one book (or bored by it), swap to another text.

B1. Book Recommendations

A good beginner's guide to R and applied statistical methods with linguistic applications is Natalia Levshina's excellent *How to do Linguistics with R: Data Exploration and Statistical Analysis*. Not being focused on linear models, this book won't give you a very detailed breakdown of the linear model framework, but instead you get introduced to many useful exploratory techniques that are not covered here.

One of the best introductions to data visualization with `ggplot2` is Kieran Healy's *Data Visualization: A Practical Introduction*. It's gorgeous.

An easy-to-intermediate text that is quite useful with respect to mixed models is Dan Mirman's *Growth Curve Analysis and Visualization Using R*. Although the later chapters are focused on how to fit polynomials to time-varying data, the earlier chapters provide a very clear introduction to mixed models.

Intermediate to advanced textbooks that go into more detail on many of the issues covered in this book are Andrew Gelman and Jennifer Hill's monumental *Data Analysis Using Regression and Multilevel/Hierarchical Models*, and the great *Mixed Effects Models and Extensions in Ecology with R*, by Zuur and colleagues. Don't shy away from the fact that the Zuur textbook is targeted at ecologists: I have often found that statistics textbooks from ecology and biology are very good reads and the mental mapping to linguistic applications is actually not all that difficult. Gelman and Hill's textbook is more focused on sociology applications. Both books go into more detail about mixed models and generalized linear models.

There is perhaps no better statistics textbook, ever, than Richard McElreath's *Statistical Rethinking*. It is a thoroughly Bayesian journey through the world of statistics, filled with metaphors, jokes, and lucid explanations. If you are new to Bayesian modeling, *Statistical Rethinking* introduces you to the framework from first principles. However, the book requires a bit more math.

I highly recommend reading some books that are exclusively focused on R to get a firm grasp of the programming language. Moreover, learning more R will greatly improve the range of datasets you can deal with, and it will save you lots of time in the long run as you become better at automatizing tasks. Perhaps one of the best starting points is Hadley Wickham and Garrett Grolemund's *R for Data Science*, which is a thoroughly "tidy" introduction to R, focusing a lot on how to wrangle with your data. On top of that, I found Norman Matloff's *The Art of R Programming* to be a delightful read on base R programming. A good follow-up to Matloff's book is Hadley Wickham's excellent *Advanced R*. Finally, I can highly recommend Patrick Burns' very witty *R Inferno* as an R programming bedtime read (if there ever was such a thing). This one is for the more poetically inclined readers.

If you are a corpus linguist or computational linguist, consider reading Julia Silge and David Robinson's *Text Mining with R: A Tidy Approach*.

If you want to learn more about data-mining techniques, I highly recommend the phenomenal *An Introduction to Statistical Learning: with Applications in R* by James, Witten, Hastie and Tibshirani.

Moving on to books that do not deal with R implementations, I can highly recommend Zoltan Dienes' *Understanding Psychology as a Science: An Introduction to Scientific and Statistical Inference*. This book should be required reading for all students in any field, as it gives crystal-clear accounts of some of the most fundamental issues in science and statistics. If you want to mix this up with some really light bedtime reads, have a look at Larry Gonick and Woollcott Smith's *The Cartoon Guide to Statistics* and Grady Klein and Alan Dabney's *The Cartoon Introduction to Statistics*. These are good for repeating the basics.

B2. Article Recommendations

For a basic introduction to statistics, statistical inference, as well as a discussion of some common pitfalls in linguistic applications of statistics, see Vasishth and Nicenboim (2016). The follow-up paper Nicenboim and Vasishth (2016) focuses more on Bayesian modeling, and I highly recommend reading it after or along with the new Vasishth et al. (2018), which also gives a nice introduction to the Bayesian modeling package brms (Bürkner, 2017) with linguistic examples. As this book has already introduced you to the lme4 package, converting to brms will be quite easy once you understand the relevant Bayesian concepts.

I *highly* recommend reading Schielzeth (2010) for a discussion of the usefulness of centering in the presence of interactions. I also recommend Zuur et al. (2010) for a great overview of regression assumptions and collinearity. Both of these are ecology/biology papers, but they are very accessibly written. Jaeger (2008) is a good discussion of logistic regression with linguistic examples.

If you have complex nonlinearities in your data (pitch trajectories, articulatory trajectories, etc.), you may want to look into generalized additive models. I collaborated with Martijn Wieling on a tutorial on this (Winter & Wieling, 2016). There also is Sóskuthy (2018) and Wieling (2018), which are good follow-up tutorials that go into more detail.

In addition, I recommend the reader to have a look at two special issues focused on data analysis that have been published in linguistic journals: first, the 2008 *Emergent Data Analysis* special issue in the *Journal of Memory and Language*; and, second, the

2018 special issue called *Emergent Data Analysis in Phonetic Sciences* in the *Journal of Phonetics* (don't worry—you don't have to be a phonetician to understand the papers presented in there).

B3. Staying Up-to-Date

The best way to stay 'up-to-date' with R and statistics is to engage with the vast online community of data scientists. People in this community are incredibly willing to share their knowledge with others. There's a plethora of free tutorials online. I can also highly recommend following data scientists and quantitative linguists on Twitter.

The most important thing is that you continue learning. I hope that this book is a stepping stone for you.

Index

Index of R Functions